Approaches to Heavy Metal Tolerance in Plants

Sumira Jan • Javid Ahmad Parray

Approaches to Heavy Metal Tolerance in Plants

Springer

Sumira Jan
Division of Biotechnology
ICAR-Central Institute of Temperate Horticulture
Srinagar, Jammu and Kashmir, India

Javid Ahmad Parray
Centre of Research for Development
University of Kashmir
Srinagar, Jammu and Kashmir, India

ISBN 978-981-10-1692-9 ISBN 978-981-10-1693-6 (eBook)
DOI 10.1007/978-981-10-1693-6

Library of Congress Control Number: 2016950721

Printed on acid-free paper

This Springer imprint is published by Springer Nature
The registered company is Springer Science+Business Media Singapore Pte Ltd.

Dr. Mahmooduzzafar

Prof. Azra N Kamili

Dedicated to our respected mentors and teachers
Thanks for infusing us with curiosity to learn nature and respect natural laws.
We don't know where your influence stops.
With you, we became seekers and explorers.

Contents

About the Authors

Sumira Jan is presently working as research scientist in ICAR – Central Institute of Temperate Horticulture at Rangreth, Srinagar, Jammu, and Kashmir, 190007, India. She obtained her M. Sc. and Ph.D. degrees from Jamia Hamdard, New Delhi. She completed her postdoc in 2014. Later on, she joined as tenure track, early career scientist at the Centre of Research for Development (CORD) at the University of Kashmir. She has been actively engaged in research and teaching in the field of plant physiology and ecophysiology since 2007. Her major research interests include plant biochemistry and plant metabolomics, radiation biology, stress ecophysiology, and medicinal and aromatic herbs along with their mass cultivation and bioefficacy. Dr. Jan has been awarded with doctoral research grant by the Korean Government Scholarship Program (KGSP) through MHRD in the field of environment. Dr. Jan is a recipient of several awards including both national and international scientific organizations. In addition to her awards, she has been awarded with numerous research grants in the field of plant sciences from UGC, DBT, and DST. Apart from research expertise, she has also advanced her research towards the biological efficacy of herbs as well as plant improvement. Her work has been published in various international journals. She has consistent flair for scientific writing and loves to communicate her research findings. She has been an active member of the American Society of Plant Physiology and Indian Women Scientists.

Javid Ahmad Parray is presently working as research scientist at the Centre of Research for Development (CORD) at the University of Kashmir, Srinagar, Jammu, and Kashmir, 190006, India. He is a recipient of the Emerging Scientist Award 2015 by the Indian Academy of Environmental Science (IAES) Haridwar. He obtained his M. Sc. M.Phil. and Ph.D. degrees from the University of Kashmir. He has qualified for state-level competitive examinations like SET. He completed his doctorate in 2014. After that, he joined as research associate in an important network mission on saffron. He has contributed numerous international papers and added various data sets pertaining to crop improvement via PGPRs. He has been actively engaged in research and teaching in the field of environment sciences, environment biotechnology, microbial ecology and plant microbe interactions since 2008. His main research subject encompasses microbial diversity, ecophysiology, stress physiology, and medicinal and aromatic herbs along with their mass cultivation and bioefficacy. Dr. Javid has presented his work at national and international platforms and was awarded with numerous awards for his immense contribution in the field of plant improvement.

Heavy Metal Uptake in Plants

1

Contents

Abstract

Metal uptake is highly a complex process involving metal transfer from soil sap inside cells of roots. The availability of metals and their motility in the rhizosphere is highly mediated by root exudates and microorganisms. Higher plants possess highly effective systems for the acquisition nutrient contents from the soil. Plants can make metal ions more available for uptake by acidifying the rhizosphere and pumping protons via plasma membrane-localized proton pumps and also by exuding low-molecular-weight (LMW) compounds that act as metal chelators.

Keywords

Metal transporters • Phytochelatins • Metallothionein • ZIP family

1.1 Introduction

Plants procure their nutrients from soil, but they are not selective enough to take up essential elements only but sometimes absorb nonessential or even toxic elements. Metal uptake is highly a complex process involving metal transfer from soil sap to inside cells of roots (Tangahu et al. 2011). Soil sap enters the root through root hairs. Saifullah et al. (2009) elaborated mechanism of metal uptake in which soil sap first enters sym-

S. Jan, J.A. Parray, *Approaches to Heavy Metal Tolerance in Plants*,
DOI 10.1007/978-981-10-1693-6_1

plast by crossing plasma membrane and then passes through cells by means of plasmodesmata. Soil sap makes its way through apoplast; apoplastic solution enters the symplasm of the endodermis cells which can later pass through plasmodesmatal connections into cells of stele. Soil solution then can move between cells and then enters into xylem. Once metal ions are taken up by roots, they are either stored by roots or transported to the shoot (Aleel 2009). Transpiration pull brings mass upward flow of water which is essential for transport of metal ions to shoots (Annamalainathan et al. 2013). Metal translocation through xylem is facilitated by organic acids and amino acids. Some essential transition elements are known as micronutrients like iron, manganese, molybdenum, copper, zinc, and nickel because they are required by plants in small quantities. Some transition metals such as silver, gold, and cobalt and non-transition elements like aluminum have shown stimulatory effect on plant growth but are not considered essential. Plants take up nutrient elements from soils along the concentration gradients which are induced selective ion uptake or by diffusion of elements. The level of accumulation of elements varies between different species (Annamalainathan et al. 2013). Plants are classified into three categories on the basis of their response to heavy metals: Plants colonizing metal-contaminated soils have developed three basic strategies (Meriem et al. 2015): (1) metal excluders, the excluder group includes the majority of plant species that limit the translocation of heavy metals and maintain low levels of contaminants in their aerial tissues over an extensive range of soil concentrations; (2) metal indicators, plants that are metal indicators accumulate metals in their harvestable biomass, and these levels generally are reflective of the metal concentration in the soil; and (3) accumulators, metal accumulators are plants that increase internal sequestration, translocation, and accumulation of metals in their harvestable biomass to levels that far exceed those found in the soil. This term was coined by Baker and Brooks (1989) to define plants that contained > 1000/g g^{-1} of Ni > 100 mg Cd kg^{-1} (0.01 %) or > 500 mg Cr kg^{-1} (0.05 %) in dry leaf tissue that can be considered as hyperaccumulators (Baker et al. 2000).

Contaminant uptake by plants and its mechanisms have been being explored by several researchers. It could be used to optimize the factors to improve the performance of plant uptake. The plants act both as "accumulators" and "excluders." Accumulators survive despite concentrating contaminants in their aerial tissues. These accumulators biodegrade or biotransform the contaminants into inert forms in their tissues. The excluders restrict contaminant uptake into their biomass (Maestri et al. 2010). Plants have evolved highly specific and very efficient mechanisms to obtain essential micronutrients from the environment, even when present at low ppm levels. Plant roots, aided by plant-produced chelating agents and plant-induced pH changes and redox reactions, are able to solubilize and take up micronutrients from very low levels in the soil, even from nearly insoluble precipitates. Plants have also evolved highly specific mechanisms to translocate and store micronutrients (Adamczk-Szabela et al. 2015). These same mechanisms are also involved in the uptake, translocation, and storage of toxic elements, whose chemical properties simulate those of essential elements. Thus, micronutrient uptake mechanisms are of great interest to phytoremediation. The range of known transport mechanisms or specialized proteins embedded in the plant cell plasma membrane involved in ion uptake and translocation includes (1) proton pumps (″-ATPases that consume energy and generate electrochemical gradients), (2) co- and antitransporters (proteins that use the electrochemical gradients generated by ″-ATPases to drive the active uptake of ions), and (3) channels (proteins that facilitate the transport of ions into the cell). Each transport mechanism is likely to take up a range of ions. A basic problem is the interaction of ionic species during uptake of various heavy metal contaminants. After uptake by roots, translocation into shoots is desirable because the harvest of root biomass is generally not feasible. Little is known regarding the forms in which metal ions are transported from the roots to the shoots (Masoud et al. 2013).

Plant uptake–translocation mechanisms are likely to be closely regulated. Plants generally do not accumulate trace elements beyond near-term metabolic needs. And these requirements are small ranging from 10 to 15 ppm of most trace elements suffice for most needs (Peijnenburg et al. 2000). The exceptions are "hyperaccumulator" plants, which can take up toxic metal ions at levels in the thousands of ppm. Another issue is the form in which toxic metal ions are stored in plants, particularly in hyperaccumulating plants, and how these plants avoid metal toxicity. Multiple mechanisms are involved. Storage in the vacuole appears to be a major one. Water, evaporating from plant leaves, serves as a pump to absorb nutrients and other soil substances into plant roots (Thomas and Anthony 2011). This process, termed evapotranspiration, is responsible for moving contamination into the plant shoots as well. Since contamination is translocated from roots to the shoots, which are harvested, contamination is removed while leaving the original soil undisturbed. Some plants that are used in phytoextraction strategies are termed "hyperaccumulators." They are plants that achieve a root-to-shoot metal–concentration ratio greater than one (Zhao et al. 2011). Nonaccumulating plants typically have a root-to-shoot ratio considerably less than one (Masoud et al. 2013). Ideally, hyperaccumulators should thrive in toxic environments, require little maintenance, and produce high biomass, although few plants perfectly fulfill these requirements. Metal accumulating plant species can concentrate heavy metals like Cd, Zn, Co, Mn, Ni, and Pb up to 100 or 1000 times those taken up by nonaccumulator (excluder) plants (Evangelou et al. 2013). In most cases, microorganism bacteria and fungi, living in the rhizosphere closely associated with plants, may contribute to mobilize metal ions, increasing the bioavailable fraction. Their role in eliminating organic contaminants is even more significant than that in case of inorganic compounds (Li et al. 2005).

1.2 Metal Ion Binding to Extracellular Exudates and to the Cell Wall

As a first line of defense against heavy metals, plant roots secrete exudates into the soil matrix. One of the major roles of root exudates is to chelate metals and to prevent their uptake inside the cells. For example, Ni-chelating histidine and citrate are present in root exudates, and these reduce the uptake of Ni from soil (Nishida et al. 2011). The binding of metal ions such as Cu and Zn in the apoplast also helps to control the metal content of root cells (Zhao et al. 2011). Cation binding sites are also present on the root cell wall, and this allows metal exchange thus influencing the availability of ions for uptake and diffusion into the apoplast (Yang et al. 2006). The cell wall can play a key role in the immobilization of toxic heavy metal ions by providing pectic sites and histidyl groups, and extracellular carbohydrates such as callose and mucilage, and thus prevents heavy metal uptake into the cytosol. Thus, different tobacco genotypes with chemically distinct root cell wall surfaces have different sensitivities to Mn toxicity. These data suggest that the chemical properties of the cell wall might modulate plant metal uptake and consequently metal tolerance. However, the role of the cell wall in metal tolerance is not completely understood (Verbruggen et al. 2009). The cell wall is in direct contact with metal ions in the soil, but only a limited number of absorption sites are available, suggesting the cell wall has only a minor impact on metal tolerance. However, *Silene vulgaris* ssp. humilis is a heavy metal-tolerant plant that accumulates different heavy metals by binding them to proteins or silicates present in their epidermal cell walls (Revathi and Subhashree 2015).

Arabidopsis thaliana has eight genes encoding members of the type 1_B heavy metal-transporting subfamily of the P-type ATPases. Three of these transporters, HMA2, HMA3, and HMA4, are closely related to each other and are most similar in sequence to the divalent heavy metal cation transporters of prokaryotes (Dorota et al. 2015). To determine the function of these transporters in metal homeostasis, we have identified and characterized mutants affected in each. Whereas the individual mutants exhibited no apparent phenotype, *hma2 hma4* double mutants had a nutritional deficiency phenotype that could be compensated for by increasing the level of Zn, but not Cu or Co, in the growth medium (Verbruggen et al. 2009). Levels of Zn, but not other essential elements, in the shoot tissues of a *hma2 hma4* double mutant and, to a lesser extent, of a *hma4* single mutant were decreased compared with the wild type. Together, these observations indicate a primary role for HMA2 and HMA4 in essential Zn homeostasis. *HMA2promoter-* and *HMA4promoter-*reporter gene constructs provide evidence that *HMA2* and *HMA4* expression is predominantly in the vascular tissues of roots, stems, and leaves. In addition, expression of the genes in developing anthers was confirmed by RT-PCR and was consistent with a male-sterile phenotype in the double mutant (Axelsen and Palmgren 1998). HMA2 appears to be localized to the plasma membrane, as indicated by protein gel blot analysis of membrane fractions using isoform-specific antibodies and by the visualization of an HMA2-green fluorescent protein fusion by confocal microscopy (Clemens 2006). These observations are consistent with a role for HMA2 and HMA4 in Zn translocation. *hma2* and *hma4* mutations both conferred increased sensitivity to Cd in a phytochelatin-deficient mutant background, suggesting that they may also influence Cd detoxification (Mehes-Smith et al. 2013; Yang and Chu 2011).

1.3 Metal Ion Transport Through the Plasma Membrane in Roots

Plants utilize various transporters for metal uptake and homeostasis. At the cellular level, plasma membrane and tonoplast metal transporters maintain physiological concentrations of heavy metals, but they may also contribute to heavy metal stress responses. These transporters belong to the heavy metal P1B-ATPase, the NRAMP, the CDF (Williams et al. 2000), and the ZIP families (Guerinot 2010). In plants, most of these metal ion transporters were identified by complementation in *Saccharomyces cerevisiae* mutants defective in metal uptake.

1.3.1 ZIP Family

ZIP transporters are identified from a number of plant species, mainly dicots, reviewed by Grotz and Guerinot (2006).[7] They are involved in transport of various metal ions, such as Mn^{2+}, Fe^{2+}/Fe^{3+}, Cd^{2+}, Co^{2+}, Cu^{2+}, Ni^{2+}, and especially Zn^{2+}. The IRT1 homologs, AtIRT1, OsIRT1, and HvIRT1 have been localized to the plasma membrane, and they have been proposed to have their main role in Fe^{2+}/Fe^{3+} and Mn2+ uptake, respectively (Vert et al. 2001). In addition, OsZIP4 has been appointed a role in Zn2+ uptake as this protein is localized to the plasma membrane, whereas GmZIP1 is specifically expressed in the peribacteroid membrane in root nodules of soybean plants. The cell- and tissue-specific localization of the remaining identified *ZIP* proteins is currently unknown, and therefore the specific role, besides being involved in metal homeostasis, awaits further studies (Nishida et al. 2008). The number of *ZIP* genes in *Arabidopsis* (15 putative ZIP proteins)7 and rice (12 putative ZIP proteins) is rather remarkable as the high number may indicate some redundancy. Varying substrate specificities and differential expression at different membranes and tissues during plant development can to some extent explain the abundancy of *ZIP* genes present in both monocot and dicot plant species. So far there have been identified five putative barley ZIP proteins, and these have been named according to nomenclature used for the rice homologous. The three ZIP proteins analyzed in the reference study seem to be Zn2+ specific, and the expression level for *HvZIP3* and *HvZIP5* is induced by a Zn deficiency treatment, whereas *HvZIP8* was constitutively expressed (Kramer et al. 2007). There are also some evidences that Ni is taken up by Zn transporters although candidate Ni-specific transporters have also been identified (Fig. 1.1) (Assunçao et al. 2001).

1.3.2 NRAMP Family

Members of the NRAMP family have been identified in all plants studied at the molecular level (Thomine et al. 2000). In *Arabidopsis*, NRAMPs were shown to transport Fe, Mn, and Cd. When expressed in yeast, AtNRAMP1, AtNRAMP3, and AtNRAMP4 mediate the uptake of Fe, Mn, and Cd (Curie et al. 2000; Thomine et al. 2000). AtNRAMP3 and AtNRAMP4 are closely related, have similar gene expression patterns and regulation by Fe, and both localize at the vacuolar membrane. AtNRAMP3 and AtNRAMP4 function redundantly to release metals from the vacuole (Cailliatte et al. 2010). These proteins are involved in the mobilization of seed vacuolar iron stored during germination and were shown to provide Mn to maintain photosynthetic activity under Mn starvation (Lanquar et al. 2005). Transcriptomic analyses highlighted the induction of AtNRAMP3 during leaf senescence, indicating that this transporter could play a role in metal remobilization during leaf senescence (Takahashi et al. 2011). AtNRAMP1 was shown to encode the high-affinity Mn uptake system in *Arabidopsis* roots (Cailliatte et al. 2010). Similarly, OsNRAMP5 allows Mn uptake and translocation in rice roots (Yang et al. 2004) and is also in the main pathway for Cd accumulation in rice (Ishikawa et al. 2012) *A. halleri* and *Thlaspi caerulescens* metal hyperaccumulating species, NRAMP3 and NRAMP4 are expressed at higher levels than their *A. thaliana* homologs (Oomen et al. 2009). In addition, TcNRAMP3 is induced by the Fe starvation and by the heavy metals Cd and Ni in roots (Wei et al. 2009). Yeast expressing TcNRAMP3 accumulated Cd and excluded Ni (Oomen et al. 2009). In the Ni-hyperaccumulating species *P. gabriellae*, PgNRAMP1.1 and PgNRAMP2.1 were characterized by expression in yeast and shown to be able to transport Fe and Mn (Curie et al. 2000) (Fig. 1.2).

1.3.3 Copper Transporter Family

Despite copper ions being crucial in proteins participating in plant processes such as electron transport, free-radical elimination, and hormone perception and signalling, very little is known about copper inward transport across plant membranes. In this work, a five-member family

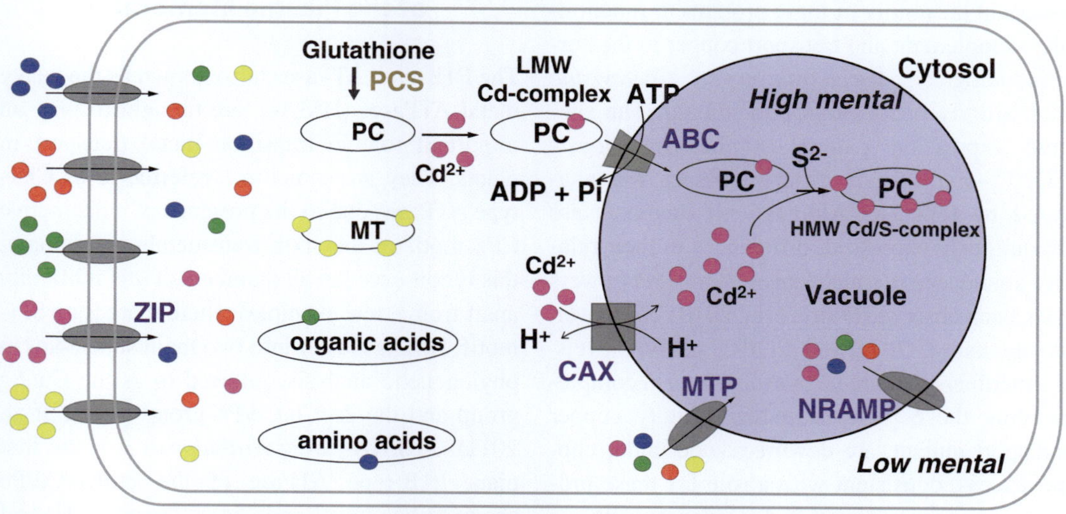

Fig. 1.1 Vacuolar sequestration of heavy metals in plant cell (Assunçao et al. 2001)

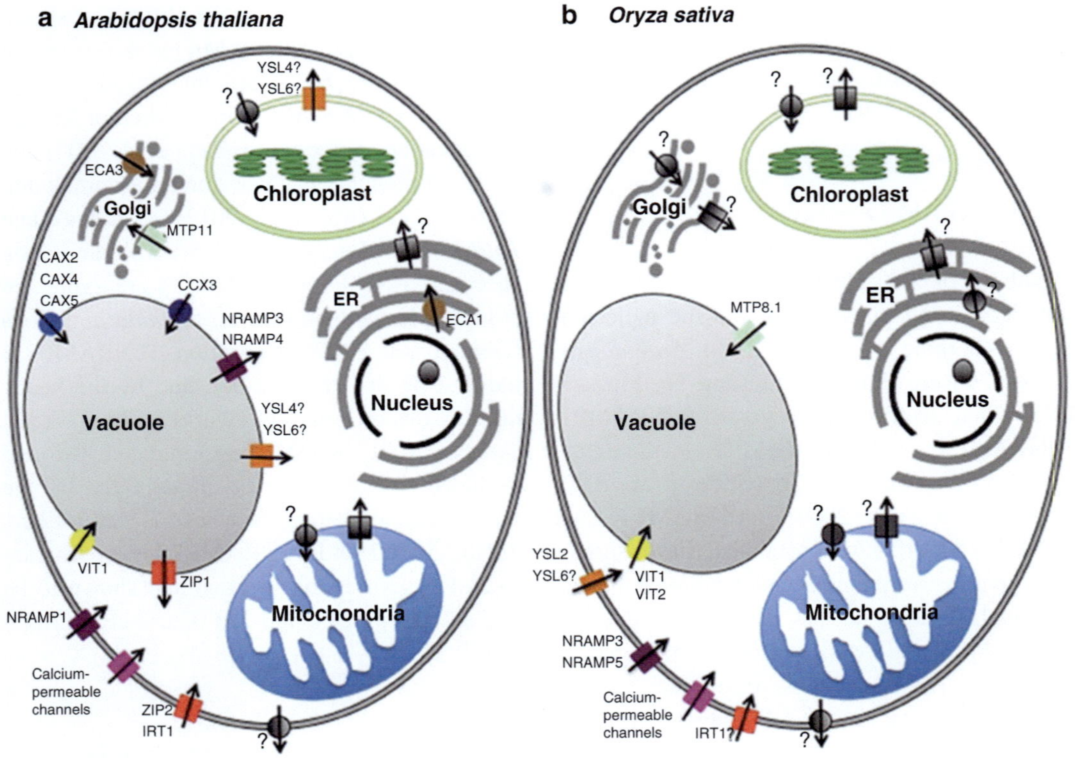

Fig. 1.2 Subcellular localization of putative Mn transporters: A diagram of a plant cell showing the Mn transport pathways in (**a**) *A. thaliana* and (**b**) *O. Sativa*. Squares, import into the cytosol; circles, export out of the cytosol; *gray*, unknown; *red*, ZIP family; *magenta*, calcium-permeable channels; *orange*, YSL family; *light blue*, CAX family; *dark Blue*, CCX family; *yellow*, VIT family; *purple*, NRAMP family; *brown*, P2A-Type ATPase family; *green*, CDF/MTP family (Source: Socha and Guerinot 2014)

(COPT1–5) of putative *Arabidopsis* copper transporters is described (Burkhead et al. 2009). We ascertain the ability of these proteins to functionally complement and transport copper in the corresponding *Saccharomyces cerevisiae* high-affinity copper transport mutant. The specific expression pattern of the *Arabidopsis* COPT1–5 mRNA in different tissues was analyzed by RT-PCR. Although all members are ubiquitously expressed, differences in their relative abundance in roots, leaves, stem, and flowers have been observed (Nuria et al. 2010). Moreover, steady-state COPT1 and COPT2 mRNA levels, the members that are most efficacious in complementing the *S. cerevisiae* high-affinity copper transport mutant, are downregulated under copper excess, consistent with a role for these proteins in copper transport in *Arabidopsis* cells.

1.4 Reduced Metal Uptake and Efflux Pumping at the Plasma Membrane

The P1B-type ATPases, also known as the heavy metal ATPases (HMAs), are thought to play an important role in transition metal transport in plants. They are sometimes referred to as CPx-type ATPases due to the presence of a distinctive CPx motif in the sixth transmembrane domain; this is conserved in all members of this subfamily apart from a few examples where there is an SPC motif. HMAs cluster into two main subclasses in phylogenetic analysis, referred to as the Cu/Ag group and the Zn/Co/Cd/Pb group (Wang et al. 2011). AtHMA4 from *Arabidopsis* was the first plant P1B-type ATPase of the Zn/Co/Cd/Pb group to be cloned and characterized. AtHMA4 confers Cd resistance when heterologously expressed in a wild-type (wt) strain of *S. cerevisiae*, and it also rescues the Zn sensitivity of the *E. coli* zntA mutant, suggesting a role in Cd and Zn transport. These metals also regulate AtHMA4 expression in *Arabidopsis*. Evidence for a role in Zn nutrition was provided recently when it was shown that an *Arabidopsis* hma2 hma4 double mutant showed a phenotype consistent with Zn deficiency and that this phenotype could be suppressed by adding Zn to the growth medium. This phenotype was not observed in the single mutants, suggesting partially redundant functions of AtHMA4 and AtHMA2 (Colangelo and Guerinot 2006). Expression studies using HMA-promoter-GUS constructs indicated that AtHMA2 and AtHMA4 were predominantly expressed in the vascular tissues of roots, stems, and leaves and that AtHMA2 was localized to the plasma membrane. It therefore appears that AtHMA2 and AtHMA4 may play a role in the translocation of Zn and may function more specifically in the loading or unloading of Zn in the xylem and in remobilization of Zn from shoot to root in the phloem. No altered sensitivity to Zn and Cd was observed for the single hma4 mutant, although the concentration range over which these metals were tested was not stated. However, in a phytochelatin-deficient mutant background, hma2 and hma4 mutations increased the sensitivity to low levels of Cd (Kim et al. 2007). A further mutant allele for AtHMA4 (athma4-3) has been isolated here, and additional phenotypes in the single hma4 mutants are reported that are consistent with possible detoxification roles for AtHMA4 when metals are supplied at high concentration. Evidence is presented using heterologous expression in yeast to show that AtHMA4 functions as an efflux pump conferring both Cd and Zn resistance (Kobae et al. 2006).

1.5 Root-to-Shoot Metal Translocation

The translocation of metals to the aerial biomass can be an important biochemical process used by plants to remediate polluted areas. In some plants, the mobilization of metals from their roots to their above aerial organs can minimize the damage that could be exerted by these heavy metals on the root physiology and biochemistry (Kramer et al. 2007) Excluders prevent or limit the translocation of toxic metals or essential metals from roots to shoots. On the other hand, accumulators/hyperaccumulators translocate metals from roots to shoots via the xylem with the transpiration stream. This is accomplished by increasing the uptake of metals in roots and by reducing the sequestration of metals in the root. The chelation of metals with ligands, such as organic acids, amino acids, and thiols facilitates the movements of heavy metals from roots to shoots (Zacchini et al. 2009). The xylem cell wall has a high cation exchange capability; thus, the movement of metal cations is severely retarded when the metals are not chelated by ligands. Organic acids are involved in the translocation of Cd in the species *Brassica juncea* (Salt and Kramer 2000). The chelation of Ni to histidine is involved in the long-distance translocation of Ni in the hyperaccumulator *A. lesbiacum*, where a 36-fold increase was reported in the histidine content of the xylem sap upon exposure to nickel (Zacchini et al. 2009). Richau et al. (2009) found that the Ni hyperaccumulator, *T. caerulescens*, had a higher free-histidine concentration in roots compared to the non-Ni hyperaccumulator *T. arvense*. Also, *T. caerulescens* had less Ni in root vacuoles than *T. arvense* because the histidine–Ni complexes were much less taken up by vacuoles than free Ni ions. Therefore, an increase in free histidine in roots inhibited the vacuolar sequestration of His–Ni in *T. caerulescens* compared to free Ni in *T. arvense* and also had enhanced histidine-mediated Ni xylem loading. The elevated free histidine in root cells appears to be involved in reduced vacuolar sequestration and enhanced xylem loading of Ni (Richau and Schat 2009). This was also the case for Zn and Cd for this hyperaccumulating species (Hassan and Aarts 2011). An increase in Ni accumulation was also observed in the Ni hyperaccumulator *Sebertia acuminata* where, when chelated to citrate, Ni was able to translocate to the shoot. In the absence of citrate, Ni was no longer accumulated in the aerial tissues (Lee et al. 1977). The chelation of metals with nicotianamine (NA) also contributes to improved tolerance. Nicotianamine can chelate and transport divalent Ni, Cu, and Zn (Takahashi et al. 2003; Pich et al. 2001; Ling et al. 1999). The nicotianamine synthase (NAS) enzyme is responsible for the synthesis of NA by trimerization of S-adenosylmethionine (Shojima et al. 1990). When exposed to high levels of Zn, Cd, and/or Ni, all four NAS genes were highly expressed in *T. caerulescens* compared to nonhyperaccumulator *A. thaliana* (van de Mortel et al. 2006). In the presence of elevated Mn, Zn, Fe, and Cu concentrations, Kim et al. (2004) reported an increased expression of the NAS gene, as well as NA levels for *A. thaliana* and *N. tubacum*. In addition, Pianelli et al. (2005) showed that the overexpression of the *T. caerulescens* NAS3 gene in the Ni excluder *A. thaliana* resulted in improved Ni tolerance and Ni accumulation in their aerial organs. An increase of Fe, Zn, and Cu accumulation in *O. sativa* was associated with an overexpression of the NAS3 gene (Hassan and Aarts 2011).

1.6 Metal Storage

Metals are first transported by the process of xylem loading. These are then detoxified in shoots and finally redistributed through phloem. Ounces unloaded metals are transferred to the surrounding cells either symplastically or apoplastically (Hassan and Aarts 2011). It is reported that yellow stripe-like proteins (YLS) play an important role in symplastic transport (DiDonato et al. 2006). In *T. caerulescens* (hyperaccumulator), three *YSL* genes (*TcYLS3*, *TcYSL5*, and *TcYSL7*) are found highly expressed in shoots around vascular tissues. However, this high level of expression is not found in the *A. thaliana* (excluder plant) (Hassan and Aarts 2011). Thus, it is suggested that *TcYSL3* has a function of unloading Ni–NA complexes from the xylem into leaf cells and to distribute them to storage cells. By yeast complementation studies, it was also determined that TcYSL3 was also a Fe/Ni–NA influx transporter (Hassan and Aarts 2011). The sequestration of excess essential and nonessential metals is localized in different parts of the aerial tissue, such as leaf epidermal cell, trichomes, and vacuole. Broadhurst et al. (2004) reported *Alyssum* hyperaccumulator species that the majority of hyperaccumulated Ni is stored in either leaf epidermal cell vacuoles or in the basal section of stellate trichomes. In *A. halleri*, the majority of Zn ions are found stored in the vacuoles of mesophyll cells, while for *T. caerulescens*, most Zn ions are found in the vacuoles of epidermal cells (Verbruggen et al. 2009).

1.6.1 HMA Family of Transporters

Although HMA genes have been extensively studied in various plants, including *Arabidopsis*, rice, and barley, investigations on the HMA gene family in *Populus* are limited. To compare the mechanisms on metal phytoremediation between woody and herbaceous plants that maintain different life cycles and considering the importance of the HMA gene family in plant responses to heavy metal stresses, we investigated *HMA* genes in *Populus* (Andres-Colas et al. 2006). In the present study, we performed a genome-wide analysis of the *P. trichocarpa* HMA gene family, its phylogenetic analysis, chromosomal distribution, and expressional analysis (Verret et al. 2004). We identified a total of 17 HMA genes in *P. trichocarpa*. Quantitative real-time RT-PCR (qRT-PCR) showed that the HMA genes in *Populus* were differentially regulated by excessive Cu, Ag, Zn, Cd, Co, Pb, and Mn stress (Mills et al. 2003). The results provide insights for future investigations into the roles of these candidate *HMA* genes in response to metal stress in *Populus*.

1.6.2 MATE Family of Efflux Proteins

Multidrug and toxic compound extrusion (MATE) transporters represent a large family in plants. Plants have a higher diversity of MATE-type transporters than bacteria and animals. For example, in *Arabidopsis*, there are 58 MATE orthologs. Recently, three studies have reported that some MATE proteins are involved in the transport of citrate, which is required for iron (Fe) translocation or aluminum (Al) detoxification. FRD3 from *Arabidopsis* has been demonstrated to be a citrate transporter, which is required for Fe translocation from the roots to the shoots (Durrett et al. 2007). It is localized at the pericycle and cells internal to the pericycle cells in roots. Defects in this transporter resulted in the precipitation of Fe in the root vasculature (Durrett et al. 2007).

1.6.3 Oligopeptide Transporter Family

Genes in the oligopeptide transporter family encode integral membrane proteins that are believed to translocate their substrates from either the extracellular environment or an organelle into the cytosol Cd (Schaaf et al. 2004). Phylogenetic analyses of plant transporters have revealed two distant clades: the yellow stripe-like (YSL) proteins and the so-called oligopeptide transporters (OPTs), for which the family was named. Three categories of substrates have been identified for this family: small peptides, secondary amino acids bound to metals, and glutathione (Curie et al. 2001). Notably, the YSL transporters are involved in metal homeostasis through the translocation of metal chelates, indicating a level of conservation both in biological function as well as substrates (DiDonato et al. 2006). In contrast, the functions of OPT proteins seem to be less defined, and, in this review, I will examine the supporting and contradictory evidence for the proposed roles of OPTs in such diverse functions as long-distance sulfur distribution, nitrogen mobilization, metal homeostasis, and heavy metal sequestration through the transport of glutathione, metal chelates, and peptides (Colangelo and Guerinot 2006).

1.7 Heavy Metal Chelation in the Cytosol

Inside the cell, heavy metal ions metabolically not required immediately are stored by diverse mechanisms so as to prevent them to reach to toxic concentrations. Specific peptides such as MTs and PCs are used to chelate metals in the cytosol of the cell and then to sequester them in specific subcellular compartments. A large number of small molecules are also involved in metal chelation inside the cells, including phosphate derivatives, amino acids, and organic acids (Andra et al. 2010).

1.7.1 Phytochelatins

One of the mechanisms adopted by plants to detoxify HMs is the production of short-chain thiol-rich repetitions of peptides of low-molecular weight synthesized from sulfur-rich glutathione (GSH) by the enzyme phytochelatin synthase (PCS) with the general structure of (γ-glutamyl-cysteinyl) n-glycine (n=2 to 11) that have a high affinity to bind to HMs when they are at toxic levels (Hassinen et al. 2011). Phytochelatins, as a pathway for metal homeostasis and detoxification, have been identified in a wide range of living organisms from yeast and fungi to many different species of animals. In plants, PCs are found to be part of the defensive act not only against metal-related stresses but also in response to other stressors such as excess heat, salt, UV-B, and herbicide. PCs are reported to be used as biomarkers for the early detection of HM stress in plants (Blindauer 2008). Cytosol is the place where PCs are manufactured and actively shipped from there in the form of metal–phytochelatin complexes of high molecular weight to vacuole as their final destination. It has been suggested that the transport is mediated by Mg ATP-dependent carrier or ATP-binding cassette (ABC) transporter (Cobbett and Goldsbrough 2002). The precipitous induction of PCs occurs inside cells as a result of the varying levels of multiple types of HMs where PCs via sulfhydryl and carboxyl groups can attach to some HM cations and anions such as Cd, Cu, Ag, Zn, Pb, Ni, and Ar. However, Cd2+ ions are found to be the most effective stimulator of PC synthesis where they are four to sixfold stronger in inducing PCs than Cu2+ and Zn2+ in cell cultures of *Rauvolfia serpentina* and red spruce (*Picea rubens* Sarg), respectively (Hassinen et al. 2011). PCs can be both produced and accumulated in roots and aerial organs. Nevertheless, the majority of studies suggest that they tend to be first biosynthesized and amass in roots. It has been shown that, in sunflower exposed to Cd intoxication, phytochelatin levels in roots were at least two times as much as those in leaves. PC chain lengths show variation within and among plant species as well as with HM types. Brunetti et al. reported that PC4 was most pervasive oligomer in tobacco seedlings (*Nicotiana tabacum* L.), whereas PC3 was of higher concentration in *Arabidopsis*. In legumes, it is reported that PCs with longer chains are stronger binder to Pb in comparison to shorter PCs. But there is no conclusive study to show whether the number of chains can have any impacts on the effectiveness of the PCs (Hassinen et al. 2011). Phytochelatins along with antioxidative enzymes can form a synergistic defensive regime in plants under HM stress which, in turn, can strengthen plant's resistance against metal intoxication. It was demonstrated that the increased enzymatic biosynthesis of PCs coupled with the heightened activity of antioxidative system in *Brassica chinensis* L. led to an effective detoxification of Cd. A considerable effort has been made to identify and clone PCS genes that are responsible for the production of PCs. *Arabidopsis thaliana* phytochelatin synthase (AtPCS1) and wheat (*Triticum aestivum* L.) phytochelatin synthase (TaPCS1) were among the first plant PCS genes that were extracted. The ongoing investigation into this area has led to the identification of various PCS genes in distinct plant species such as *Brassica juncea* (BjPCS1) and rice (*Oryza sativa* L.) (OsPCS1) (Cobbett and Goldsbrough 2002). Real or synthetic expression of these genes in PC deficient and transgenic plants or hyperaccumulators offers a very promising future for the possibility of increasing plant resistance against HMs and also phytoremediation strategies. In transgenic tobacco plants, artificial synthesis of phytochelatin gene enhanced their resistance to varying levels of cadmium (Anjum et al. 2012). It was showed that transgenic *Arabidopsis* plants were much better HM accumulators than wild-type *Arabidopsis* as a result of expressing synthetic phytochelatins (ECs) (Song et al. 2010). Guo et al. demonstrated that overexpression of arsenic-phytochelatin synthase 1 (AsPCS1) and yeast cadmium factor 1 (YCF1) (isolated from garlic and baking yeast) in *Arabidopsis thaliana* resulted in an increased tolerance to Cd and As and also enhanced its ability to accumulate the metals to a greater extent (Suresh and Ravishankar 2004).

1.7.2 Metallothioneins (MTs)

MTs are another family of small cysteine-rich, low-molecular-weight cytoplasmic metal-binding proteins or polypeptides that are found in a wide variety of eukaryotic organisms including fungi, invertebrates, mammals, and plants as well as some prokaryotes proteins (Cobbett and Goldsbrough 2002). Contrary to PCs that are the product of enzymatically synthesized peptides, MTs are synthesized as a result of mRNA translation. Whereas PCs in plants may mainly deal with Cd detoxification, MTs appear to be capable of showing affinity with a greater range of metals such as Cu, Zn, Cd, and As (Pomponi et al. 2006). MTs exhibit different characteristics and functionality based on their occurrence in different organisms; however, as our understanding toward the roles of plant MTs increases and given the fact that plant MTs are exceedingly varied in terms of their molecular properties and structural features, they are likely to have more and diverse functions in plant than any other living organisms. In plants, these ligands are involved in nullifying toxicity of HMs through cellular sequestration, homeostasis of intracellular metal ions, and metal transport adjustment (Hazrat et al. 2013). In addition to their role in HM detoxification, MTs are known to be active agents in a number of cellular-related events including ROS scavenger, maintenance of the redox level, repair of plasma membrane, cell proliferation, and its growth and repair of damaged DNA. There are a myriad of different endogenous and exogenous factors other than HMs that are able to induce the production and expression of MTs. Of these, osmotic stress, drought, extreme temperatures, nutrient deficiency, release of various hormones, natural and dark-induced tissue senescence, injuries, and viral infections can be mentioned. Plants have multiple MT types that are generally divided into four distinct subgroups according to the arrangement of Cys residues (Xiang et al. 2001). They demonstrate patterns of organ and developmental stage specificity so that type 1 MTs are mainly expressed in roots, while the expression of type 2 MTs mostly occurs in shoots, type 3 MTs are induced in leaves and during fruit ripening, and type 4 MTs are abundant in the developing seeds. Regarding high level of sequence diversity of plant MT, each MT subgroup (MT1 to MT4) is further subdivided and referred to as isoforms. Cobbett and Goldsbrough (2002) subdivided sugarcane MT2 into three subclasses and termed them as MT2-1, MT2-2, and MT2-3, or in *Arabidopsis* MT4 is subdivided into MT4a and MT4b (Ngu et al. 2010). It seems that all four types of MTs and their isoforms identified in plants are able to bind to HMs and act as metal chelators or storehouse; however, mounting evidence suggests that on the one hand plant MTs show distinct treatment toward varying types of metals and on the other hand functionality of these plant MTs and their metal-binding and metal-affinity characteristics as well as tissue localization might be different within a plant species or among species. Song et al. (2010) reported that, in *Arabidopsis*, there is every likelihood that MT isoforms from types 1 and 2 (1a, 2a, and 2b) and 3 are involved in copper chelation, while MT isoforms from type 4 (4a and 4b) act as a zinc binder (Xue et al. 2009). In some mutants of *Arabidopsis*, MT1 may play a more important role in detoxifying copper in leaf veins than in leaf mesophyll. Xiang et al. (2001) showed that the induction of OsMT1a (*Oryza sativa* L. metallothionein type 1) was crucial to the zinc homeostasis in roots of rice. In grain-filling and mature seeds of barely, it was demonstrated that the primary function of MT3 is to maintain homeostasis of Zn and Cu, whereas MT4 was involved in storage of Zn (Zhao et al. 2010). In soybean, it was shown that MT1, MT2, and MT3 were more likely to get involved in detoxification of deleterious amounts of Cd, while MT4 exhibited Zn-binding characteristics. It can be suggested that varying types of MTs and their isoforms have distinct and overlapping functions in homeostasis and HM detoxification (Bell and Vallee 2009). More work still needs to be done to find out the possible reasons for these differential and preferential behaviors of plant MTs toward metals; nevertheless, it appears that differences in genetic structure of plants, complex diversity in the metal-binding regions of plant MTs, and different sequence and performance of isoforms might be able to provide some explanation to the observed patterns (Song et al. 2010). Overexpression experiments are very popular with plant MTs, and expressing as well as engineering them through DNA recombinant methods into plants, yeasts, and bacteria that lack some of these proteins can increase our knowledge of MTs and their performance to a greater extent and also provide unique opportunities for phytoremediation or bioremediation strategies (Xue et al. 2009). Some works are suggestive of MTs promoting the capability of transgenic plants in terms of decreasing the production of reactive oxygen species and fortifying cellular antioxidant defense system when it comes to detoxifying excessive levels of HMs. Xia et al. showed that expression of *Elsholtzia haichowensis* metallothionein type 1 (EhMT1) in tobacco plants not only increased the tolerance of transgenic tobacco to copper toxicity but also decreased the synthesis of hydrogen peroxide and improved peroxidase activity (POD) in roots, leading to enhanced ability of plants to cope with oxidative stress. Zhou et al. demonstrated that although TaMT3, a metallothionein type 3 from *Tamarix androssowii*, engineered into tobacco resulted in increased tolerance to Cd stress through significant increases of SOD functionality, which raised the ability of ROS cleaning-up in transgenic plant, it led to decreased POD activity (Falcon and Dell 2007). It seems that the impact of the expressed metallothionein on distinct components of antioxidant system of transgenic plants is different, which requires further investigation. Ectopically expressed MTs in transgenic plants are shown to enhance their tolerance toward metal intoxication. Kumar et al. showed that OSMT1e-p, a type 1 MT obtained from a salt tolerant rice genotype (*Oryza sativa* L. cv. Pokkali), imparted tolerance toward copper and zinc toxicity when ectopically expressed in transgenic tobacco. They observed that tobacco plants that had received the gene tended to retain excessive amounts of Cu2+ and Zn+2 in their roots or lower leaves, significantly reducing the HM ion movement and content to/in upper foliage and harvestable organs. Zhao et al. (2010) concluded that the ectopic expression of BjMT2, a metallothionein type 2 from Brassica juncea, in *Arabidopsis thaliana* increased copper and cadmium tolerance at the seedling stage but acutely reduced root development when there was no heavy metal exposure. These trends may suggest that ectopic expression of MTs in transgenic plants may act in host plant in a nonspecific way and differently impact the organ growth (Oomen et al. 2009).

1.7.3 Ferritins

Ferritin is a protein occurring in almost all living organisms (except yeasts) in a very conservative spherical structure. In higher plants and animals, even the primary sequence is conserved. The molecule is the protein apoferritin coat, often called shell, filled with an iron core (Hsiao et al. 2013). The coat is usually formed from 24 structurally equivalent polypeptide subunits with molecular weight of about ~20 kDa in mammals and ~25–28 kDa in plants. Mini-ferritin, composed of 12 subunits, is present in bacteria (Ravet et al. 2009). The role of ferritin in cells is to provide iron concentration at the level necessary to their proper functioning (10–3–10–5 M). This unique molecule has the ability to accumulate iron up to the 10–2 M, while the solubility of Fe(III) under physiological conditions of pH, temperature, and the presence of air is only 10–18 M. Usually, the iron content in the molecule is below 3000 atoms, even if a higher accumulation may occur (up to 4500 atoms). Such a high concentration of iron ions in ferritin is possible as a result of the high density of the iron core, 2.5 times higher than that of the polypeptide shell (De Domenico et al. 2009). The volume of the core is only one quarter of the whole molecule. Then, the molecular weight of the protein, which for apoferritin is ~480 kDa, is almost doubled when it is filled with iron. Phytoferritin also plays an important role in the defense against pathogen attacks. Overexpression of ferritin as a result of biotic stress induces complexation of iron circulating in host organism—iron becomes inaccessible to pathogens' life processes (Hsiao et al. 2013).

1.8 Organic Acids, Amino Acids, and Phosphate Derivatives

Organic acids and amino acids bind heavy metals to heavy metals and thus can be deployed in metal toxicity. However, there is no clear correlation between heavy metal accumulation and the production of these compounds. It is reported that malate, citrate, and oxalate confer metal tolerance by transporting metals through the xylem and then sequestrating ions to the vacuoles (Gill et al. 2014). Citrate, synthesized in plants by enzyme citrate synthase, has a higher capacity for metal ions than malate and oxalate, and although its main role is to chelate Fe2+ ions, it also has a strong affinity also for Ni2+ and Cd2+. Amino acids and their derivatives are able to chelate metal ions and thus confer resistance to metal toxicity in plants. Histidine is considered as one of the most important free amino acids in heavy metal metabolism. Due to the presence of carboxyl, amino, and imidazole groups, it acts versatile metal chelator, which provides Ni tolerance and increase Ni transport in

(Kerkeb and Kramer 2003). Phytate (myo-inositol hexakisphosphate) is the main form of phosphorous reserve in plants and is often found in seeds and roots. The molecule consists of six phosphate groups which chelates various cations, like Ca2+, Mg2+, and K+ and also Fe2+, Zn2+, and Mn2+. This unique distribution of phytate and its ability to chelate multiple metal ions can be mobilized as a detoxification agent. Further, the addition of Zn to the culture medium leads to the production of Zn2+-containing phytate globoids. Thus, mobilization of phytate in these layers of cells plays a key role in metal ion loading to the aerial parts of plants. Citrate has been reported the major ligand for Ni ions as it leads in the formation of Nickel-citrate complex in *Sebertia acuminata* (Ni-hyperaccumulating plant). Nicotianamine (NA), synthesized from S-adenosyl-L L-methionine by nicotianamine synthase (NAS), has been primarily associated with Fe and Cu homeostasis (Clemens and Persoh 2009). Other amino acid chelators including glutathione, polyamines, proline, etc., are found to play roles in metal binding, metal hyperaccumulation, metal stress defense as well as signalling and antioxidation (Robinson and Winge 2010).

1.9 Metal Sequestration in the Vacuole by Tonoplast Transporters

The vacuole is a pivotal organelle functioning in storage of metabolites, mineral nutrients, and toxicants in higher plants. Accumulating evidence indicates that in addition to its storage role, the vacuole contributes essentially to long-distance transport of metals, through the modulation of vacuolar sequestration capacity (VSC) which is shown to be primarily controlled by cytosolic metal chelators and tonoplast-localized transporters or the interaction between them (Clemens and Persoh 2009). Plants adapt to their environments by dynamic regulation of VSC for specific metals and hence targeting metals to specific tissues. Study of VSC provides not only a new angle to understand the long-distance root-to-shoot transport of minerals in plants but also an efficient way to biofortify essential mineral nutrients or to phytoremediate nonessential metal pollution (Brunetti et al. 2011). The current review will focus on the most recent proceedings on the interaction mechanisms between VSC regulation and long-distance metal transport (Barry et al. 2011).

1.9.1 The ABC Transporters

Xenobiotics and heavy metals are transported by ABC transporters inside vacuoles, and it is reported that two subfamilies (MRP and PDR) are mainly active in the chelated heavy metals sequestration. Cell vacuoles are the main sites for accumulation and storage of PC–Cd complexes. PC–Cd complexes are produced within cytosol and further translocated by ABC transporters to the vacuoles (Brunetti et al. 2011). In the vacuoles, HMW complexes are formed by more Cd and sulfide incorporation. The HMW complexes are the main Cd storage form. The HMT1 is the first vacuolar ABC transporter which was identified by its ability to complement a *S. pombe* mutant. This mutant is not able to produce HMW complexes. HMT1 is located within tonoplast t, and it transports PC–Cd complexes inside the vacuole in a Mg-ATP-dependent manner. In *S. cerevisiae*, the tonoplast YCF1ABC pump transports Cd inside the vacuole as a bis(glutathionato)–Cd complex and thus provides Cd tolerance.

1.9.2 The CDF Transporters

Cation diffusion facilitators (CDFs), also known as the metal tolerance protein (MTP) family, are found in diverse groups of organisms such as bacteria, fungi, animals, and plants. Members of this family are involved in cellular heavy metal homeostasis with principal selectivity for Zn2+, Mn2+, and Fe2+ (Kobae et al. 2004). Despite their specificities for Zn2+, Mn2+, and Fe2+, many CDFs may also transport other heavy metals such as Co2+, Ni2+, and Cd2+. CDF transporters are involved in heavy metal efflux from the cytoplasm either to the extracellular space or into the organelles. Twelve *MTP* genes have been recognized so far in *A. thaliana* and ten in *O. sativa*. In *A. thaliana*, the first *CDF* gene was characterized as the *zinc transporter 1* gene (*ZAT1*) and later renamed as *metal tolerance protein 1 (AtMTP1)*. The *AtMTP1* gene is expressed constitutively in roots as well as in shoots, and when overexpressed in *Arabidopsis*, it enhances Zn tolerance (Kramer et al. 2007). However, RNA interference (RNAi)-mediated silencing or T-DNA insertion mutation of this gene increases Zn sensitivity, indicating its important role in regulation of Zn homeostasis. In *A. halleri*, a Zn hyperaccumulator plant, *AhMTP1* gene is believed to have a role in Zn hypertolerance (Goswami et al. 2012). Unlike *AtMTP1* gene, *AtMTP3* is expressed predominantly in roots and reported to be engaged in maintenance of Zn homeostasis by excluding Zn under Zn oversupply (Montanini et al. 2007). Another member of MTP family, *AtMTP11*, has been reported to transport as well as provide Mn tolerance. In rice, an ortholog of MTPs, *OsMTP1*, has been characterized and supposed to be located on chromosome 5 and highly expressed in mature leaves and stem (Sun et al. 2010).

1.9.3 The HMA Transporters

Heavy metal ATPases, *HMAs*, are found in a wide range of organisms ranging from prokaryotes to eukaryotes including yeasts, insect, mammals, and plants. Being energized by ATP hydrolysis, they translocate heavy metals (Zn, Co, Cu, Cd, and Pb) out of cytoplasm (to plasma membrane and into vacuole) (Hussain et al. 2004) and thus play important roles in their transport, compartmentalization, and detoxification. HMA members (HMA2, HMA3, and HMA4) export Zn and Cd. For instance, HMA4 protein that plays a role in nutrition and transport of Zn from root to shoot also protects plants from Cd via its efflux. Hussain et al. (2004) demonstrated that though HMA2 and HMA4 are essential for Zn homeostasis in *Arabidopsis*, double mutants of HMA2 and HMA4 (*hma2* and *hma4*) exhibited increased sensitivity of plant to Cd, suggesting that they may also play a role in Cd detoxification. Similarly, loss of function in HMA2 and HMA4 has been shown to increase Cd sensitivity in *Arabidopsis* under phytochelatins deficient, *cad1-3* as well as CAD1 backgrounds (Baxter et al. 2003). A recent study on 349 wild varieties of *A. thaliana* with combined approach of genome-wide association mapping, linkage mapping, and transgenic complementation revealed that HMA3 is solely responsible for variation in amount of Cd accumulated. Varieties with high Cd accumulation indicate toward reduced HMA3 function (Donner et al. 2012). Similarly, Song et al. (2014) found expression of C-type ATP-binding cassette (ABC) transporter (OsABCC) family in *Oryza sativa* and reported its involvement in detoxifying and reducing As accumulation in grains. They reported higher expression of these transporters under higher level of As.

1.9.4 CaCA Transporters

Control of ion concentrations is critical to cellular function. Such ion homeostasis is dependent on transporters, including ion-coupled transporters like the Na+/Ca2+ exchanger (NCX), Na+/Ca2+, K+ exchanger (NCKX), cation/Ca2+ exchanger (CCX), and H+/cation exchanger (CAX), which are members of the CaCA superfamily (Hirschi et al. 2000). CaCAs are abundant throughout life; CAX and the prokaryotic-specific YRBG-type exchangers are abundant in bacteria, while NCX, NCKX, and CCX genes are abundant in animals. In the land plant lineage, NCKX and NCX have been lost, but other types, including CAX, CCX, and MHX (Mg2+/H+ exchanger), are present within all plants. Many plant CaCAs have been functionally characterized and have key roles in cation transport and homeostasis (Sanders et al. 2002). Recently, Emery et al. 2012 performed phylogenetic analysis of CaCA genes from rice and *Arabidopsis* and argued that all genes should be classified as members of the NCX family due to the ubiquitous presence of the so-called "NCX domain" and tight phylogenetic relationship.

1.9.5 NRAMP Transporters

NRAMP is a highly conserved family of integral membrane proteins that are conserved in different species and located in the plasma membrane of root apical cells (Thomine et al. 2003). They are involved in proton-coupled active transport of various heavy metals (Fe2+, Zn2+, Mn2+, Co2+, Cd2+, Cu2+, Ni2+, and Pb2+) in wide range of organisms including bacteria, fungi, animals, and plants (Hall and Williams 2003; Cailliatte et al. 2009). However, the physiological role of NRAMP was primarily related with Fe and to a lesser extent for Mn transport (Cailliatte et al. 2009). AtNRAMP1 and 6 form the first group, and AtNRAMP2–5 constitute the second group (Oomen et al. 2009). Of these, *AtNRAMP1*, *3*, *4*, and *6* have been shown to encode functional plant heavy metal transporters (Kramer et al. 2007). Yeast mutants defective in heavy metal uptake have been utilized to investigate transport specificities of plant NRAMPs. The study revealed that *AtNRAMP1* can complement the *fet3 fet4* yeast mutant that is defective in both low- and high-affinity Fe transporters, whereas overexpression of *AtNRAMP1* in *Arabidopsis* increases plant resistance to toxic Fe concentrations (Curie et al. 2000). Furthermore, AtNRAMP3 and AtNRAMP4 have been shown to mediate the remobilization of Fe from the vacuolar store and are essential for seed germination under low Fe conditions (Thomine et al. 2003), indicating a role of NRAMPs in Fe homeostasis.

References

Adamczyk-Szabela D, Markiewicz J, Wolf WM (2015) Heavy metal uptake by herbs. IV. Influence of soil pH on the content of heavy metals in *Valeriana officinalis* L. Water Air Soil Pollut 226(4):106

Aleel K (2009) Grennan identification of genes involved in metal transport in plants. Plant Physiol 149(4):1623–1624

Andra SS, Datta R, Sarkar D, Makris KC, Mullens CP, Sahi SV (2010) Synthesis of phytochelatins in vetiver grass upon lead exposure in the presence of phosphorus. Plant Soil 326:171–185

Andres-Colas N, Sancenon V, Rodriguez-Navarro S, Mayo S, Thiele DJ, Ecker JR, Puig S, Penarrubia L (2006) The Arabidopsis heavy metal P-type ATPase HMA5 interacts with metallochaperones and functions in copper detoxification of roots. Plant J 4:225–236

Anjum NA, Ahamd I, Mohmood I, Pacheco M, Duarte AC, Pereira E (2012) Modulation of glutathione and its related enzymes in plants' responses to toxic metals and metalloids–a review. Environ Exp Bot 75:307–324

Annamalainathan K, Joseph J, Alam B, Satheesh PR, Jacob J (2013) Seasonal changes in xylem sap flow rate in mature rubber plants. J Plant Crops 41(3):343–349

Assunção AGL, Martins PD, De Folter S, Vooijs R, Schat H, Aarts MGM (2001) Elevated expression of metal transporter genes in three accessions of the metal hyperaccumulator Thlaspi caerulescens. Plant Cell Environ 24:217–226. doi:10.1111/j.1365-3040.2001.00666.x

Axelsen KB, Palmgren MG (1998) Evolution of substrate specificities in the P-type ATPase superfamily. J Mol Evol 46:84–101

Baker Alan JM, Brooks RR (1989) Terrestrial higher plants which hyperaccumulate metallic elements—a review of their distribution, ecology and phytochemistry. Biorecovery 1:81–126

Baker AJM, McGrath SP, Reeves RD et al (2000) Metal hyperaccumulator plants: a review of the ecology and physiology of a biological resource for phytoremediation of metal-polluted soils. In: Terry N, Banuelos G (eds) Phytoremediation of contaminated soil and water. Lewis Publishers, Boca Raton, pp 85–108

Barry AN, Otoikhian A, Bhatt S, Shinde U, Tsivkovskii R, Blackburn NJ, Lutsenko S (2011) The lumenal loop Met672-Pro707 of copper-transporting ATPase ATP7A binds metals and facilitates copper release from the intramembrane sites. J Biol Chem 286(30):26585–26594

Baxter I, Tchieu J, Sussman MR, Boutry M, Palmgren MG, Gribskov M, Harper JF, Axelsen KB (2003) Genomic comparison of p-type ATPase ion pumps in Arabidopsis and rice. Plant Physiol 132:618–628

Bell SG, Vallee BL (2009) The metallothionein/thionein system: an oxidoreductive metabolic zinc link. Chembiochem 10(1):55–62

Blindauer CA (2008) Metallothioneins with unusual residues: histidines as modulators of zinc affinity and reactivity. J Inorg Biochem 102:507–521

Broadhurst CL, Chaney RL, Angle JS, Maugel TK, Erbe EF, Murphy CA (2004) Simultaneous hyperaccumulation of nickel, manganese, and calcium in Alyssum leaf trichomes. Environ Sci Technol 38:5797–5802

Brunetti P, Zanella L, Proia A, De Paolis A, Falasca G, Altamura MM, Sanità di Toppi L, Costantino P, Cardarelli M (2011) Cadmium tolerance and phytochelatin content of Arabidopsis seedlings overexpressing the phytochelatin synthase gene AtPCS1. J Exp Bot 62:5509–5519

Burkhead JL, Reynolds KA, Abdel-Ghany SE, Cohu CM, Pilon M (2009) Copper homeostasis. New Phytol 182:799–816

Cailliatte R, Lapeyre B, Briat JF, Mari S, Curie C (2009) The NRAMP6 metal transporter contributes to cadmium toxicity. Biochem J 422:217–228. doi:10.1042/BJ20090655

Cailliatte R, Schikora A, Briat JF, Mari S, Curie C (2010) High-affinity manganese uptake by the metal transporter NRAMP1 is essential for Arabidopsis growth in low manganese conditions. Plant Cell 22:904–917

Clemens S (2006) Toxic metal accumulation, responses to exposure and mechanisms of tolerance in plants. Biochimie 88:1707–1719

Clemens S, Persoh D (2009) Multi-tasking phytochelatin synthases. Plant Sci 177:266–271

Cobbett C, Goldsbrough P (2002) Phytochelatins and metallothioneins: roles in heavy metal detoxification and homeostasis. Annu Rev Plant Biol 53:159–182

Colangelo EP, Guerinot ML (2006) Put the metal to the petal: metal uptake and transport throughout plants. Curr Opin Plant Biol 9:322–330

Curie C, Alonso JM, Jean MLE, Ecker JR, Briat JF (2000) Involvement of NRAMP1 from *Arabidopsis thaliana* in iron transport. Biochem J 347:749–755

Curie C, Panaviene Z, Loulergue C, Dellaporta SL, Briat JF, Walker EL (2001) Maize yellow stripe1 encodes a membrane protein directly involved in Fe(III) uptake. Nature 409:346–349

De Domenico I, Ward DM, Kaplan J (2009) Specific iron chelators determine the route of ferritin degradation. Blood 114:4546–4551

DiDonato LN, Sullivan SA, Methe BA, Nevin KP, England R, Lovley DR (2006) Role of RelGsu in stress response and Fe(III) reduction in Geobacter sulfurreducens. J Bacteriol 188:8469–8478

Donner E, Punshon T, Guerinot ML, Lombi E (2012) Functional characterisation of metal(loid) processes in planta through the integration of synchrotron techniques and plant molecular biology. Anal Bioanal Chem 402:3287–3298

Dorota AS, Justyna M, Wojciech MW (2015) Heavy metal uptake by herbs. IV. Influence of soil pH on the content of heavy metals in Valeriana officinalis L. Water Air Soil Pollut 226(4):106

Durrett TP, Gassmann W, Rogers EE (2007) The FRD3-mediated efflux of citrate into the root vasculature is necessary for efficient iron translocation. Plant Physiol 144:197–205

Emery L, Whelan S, Hirschi KD, Pittman JK (2012) Protein phylogenetic analysis of Ca2+/cation antiporters and insights into their evolution in plants. Front Plant Sci 3:245–256

Evangelou M, Brett WH, Robinson H, Madeleine S, Goerg G, Schulin R (2013) Metal uptake and allocation in trees grown on contaminated land: implications for biomass production. Int J Phytoremediation 15(1):77–90

Falcón-Pérez JM, Dell Angelica EC (2007) Zinc transporter 2 (SLC30A2) can suppress the vesicular zinc defect of adaptor protein 3-depleted fibroblasts by promoting zinc accumulation in lysosomes. Exp Cell Res 313:1473–1483

Gill SS, Gill R, Anjum NA (2014) Target osmoprotectants for abiotic stress tolerance in crop plants—glycine betaine and proline. In: Plant adaptation to environmental change: significance of amino acids and their derivatives. CAB International, Wallingford, pp 97–108

Goswami D, Kaur J, Surade S, Grell E, Michel H (2012) Heterologous production and functional and thermodynamic characterization of cation diffusion facilitator (CDF) transporters of mesophilic and hyperthermophilic origin. Biol Chem 393(7):617–629

Grotz N, Guerinot ML (2006) Molecular aspects of Cu, Fe and Zn homeostasis in plants. Biochim Biophys Acta 1763:595–608

Guerinot ML (2010) Iron in cell biology of metals and nutrients. In: Hell R, Mendel R (eds) Plant cell monographs, vol 17. Springer, Berlin/Heidelberg, pp 75–94

Hall JL, Williams LE (2003) Transition metal transporters in plants. J Exp Bot 54:2601–2613

Hassan Z, Aarts MGM (2011) Opportunities and feasibilities for biotechnological improvement of Zn, Cd or Ni tolerance and accumulation in plants. Environ Exp Bot 72:53–63

Hassinen VH, Tervahauta AI, Schat H, Kärenlampi SO (2011) Plant metallothioneins–metal chelators with ROS scavenging activity? Plant Biol 13:225–232

Hazrat A, Ezzat K, Muhammad AS (2013) Phytoremediation of heavy metals—concepts and applications. Chemosphere 91(7):869–881

Hirschi KD, Korenkov VD, Wilganowski NL, Wagner GJ (2000) Expression of *Arabidopsis*CAX2 in tobacco. Altered metal accumulation and increased manganese tolerance. Plant Physiol 124:125–134

Hsiao C, Chou IC, Okafor CD, Bowman JC, O'Neill EB, Athavale SS, Petrov AS, Hud NV, Wartell RM, Harvey SC, Williams LD (2013) RNA with iron(II) as a cofactor catalyses electron transfer. Nat Chem 5:525–528

Hussain F, Sher H, Ibrar M (2004) Ethnobotanical profile of some plants of District Swat, Pakistan. Pak J Pl Sci 10:85–104

Ishikawa S, Ishimaru Y, Igura M, Kuramata M, Abe T, Senoura T et al (2012) Ion-beam irradiation, gene identification, and marker-assisted breeding in the development of low-cadmium rice. Proc Natl Acad Sci U.S.A. 109: 19166–19171

Ishimaru Y, Takahashi R, Bashir K (2012) Characterizing the role of rice NRAMP5 in manganese, iron and cadmium transport. Sci Rep 2:286–296

Kerkeb L, Krämer U (2003) The role of free histidine in xylem loading of nickel in *Alyssum lesbiacum* and *Brassica juncea*. Plant Physiol 131:716–724

Kim D, Gustin JL, Lahner B, Persans MW, Baek D, Yun DJ, Salt DE (2004) The plant CDF family member TgMTP1 from the Ni/Zn hyperaccumulator Thlaspi goesingense acts to enhance efflux of Zn at the plasma membrane when expressed in Saccharomyces cerevisiae. Plant J 39(2): 237–251

Kim DY, Bovet L, Maeshima M, Martinoia E, Lee Y (2007) The ABC transporter AtPDR8 is a cadmium extrusion pump conferring heavy metal resistance. Plant J 50:207–218

Kobae Y, Uemura T, Sato MH, Ohnishi M, Mimura T, Nakagawa T, Maeshima M (2004) Zinc transporter of Arabidopsis thaliana AtMTP1 is localized to vacuolar membranes and implicated in zinc homeostasis. Plant Cell Physiol 45(12):1749–1758

Kobae Y, Sekino T, Yoshioka H, Nakagawa T, Martinoia E, Maeshima M (2006) Loss of AtPDR8, a plasma membrane ABC transporter of Arabidopsis thaliana, causes hypersensitive cell death upon pathogen infection. Plant Cell Physiol 47:309–318

Kramer U, Talke IN, Hanikenne M (2007) Transition metal transport. FEBS Lett 581:2263–2272

Lanquar V, Lelièvre F, Bolte S, Hamès C, Alcon C, Neumann D et al (2005) Mobilization of vacuolar iron by AtNRAMP3 and AtNRAMP4 is essential for seed germination on low iron. EMBO J 24:4041–4051. doi:10.1038/sj.emboj.7600864

Li TQ, Yang XE, He ZL, Yang JY (2005) Root morphology and Zn2+uptake kinetics of the Zn hyperaccumulator of *Sedum alfredii* Hance. J Integr Plant Biol 47(8):927–934

Maestri E, Marmiroli M, Visioli G, Marmiroli N (2010) Metal tolerance and hyperaccumulation: costs and trade-offs between traits and environment. Environ Exp Bot 68:1–13

Masoud M, Ardestani A, Cornelis AM, Gestel V (2013) Dynamic bioavailability of copper in soil estimated by uptake and elimination kinetics in the springtail Folsomia candida. Ecotoxicology 22(2):308–318

Mehes-Smith M, Nkongolo K, Cholewa E (2013) Coping mechanisms of plants to metal contaminated soil. In: Silvern S (ed) Environmental change and sustainability. InTech. doi:10.5772/55124

Meriem L, Bouamar B, Hassan EH, Abdelhak B (2015) Phytoremediation mechanisms of heavy metal contaminated soils: a review. Open J Ecol 5:375–388

Mills RF, Francini A, Ferreira da Rocha PSC, Baccarini PJ, Aylett M, Krijger GC, Williams LE (2005) The plant P1B-type ATPase AtHMA4 transports Zn and Cd and plays a role in detoxification of transition metals supplied at elevated levels. FEBS Lett 579:783–791

Montanini B, Blaudez D, Jeandroz S, Sanders D, Chalot M (2007) Phylogenetic and functional analysis of the Cation Diffusion Facilitator (CDF) family: improved signature and prediction of substrate specificity. BMC Genomics 8:107

Ngu TT, Dryden MD, Stillman MJ (2010) Arsenic transfer between metallothionein proteins at physiological pH. Biochem Biophys Res Commun 401:69–74

Nishida S, Mizuno T, Obata H (2008) Involvement of histidine-rich domain of ZIP family transporter TjZNT1 in metal ion specificity. Plant Physiol Biochem 46:601–606

Nishida S, Tsuzuki C, Kato A, Aisu A, Yoshida J, Mizuno T (2011) AtIRT1, the primary iron uptake rransporter in the root, mediates excess nickel accumulation in Arabidopsis thaliana. Plant Cell Physiol 52:1433–1442

Nuria AC, Ana PG, Sergi P, Lola PA (2010) Deregulated copper transport affects arabidopsis development especially in the absence of environmental cycles. Plant Physiol 153:170–184

Oomen RJ, Wu J, Lelievre F, Blanchet S, Richaud P, Barbier-Brygoo H, Aarts MG, Thomine S (2009) Functional characterization of NRAMP3 and NRAMP4 from the metal hyperaccumulator Thlaspi caerulescens. New Phytol 181:637–650

Peijnenburg W, Baerselman R, de Groot A, Jager T, Leenders D, Posthuma L, Van Veen R (2000) Quantification of metal bioavailability for lettuce (Lactuca sativa L.) in field soils. Arch Environ Contam Toxicol 39:420–430

Pianelli K, Mari S, Marquès L, Lebrun M, Czernic P (2005) Nicotianamine over-accumulation confers resistance to nickel in *Arabidopsis thaliana*. Transgenic Res 14:739–748

Pomponi M, Censi V, Di Girolamo V (2006) Overexpression of Arabidopsis phytochelatin synthase in tobacco plants enhances Cd2+ tolerance and accumulation but not translocation to the shoot. Planta 223:180–190

Ravet K, Touraine B, Boucherez J, Briat JF, Gaymard F, Cellier F (2009) Ferritins control interaction between iron homeostasis and oxidative stress in Arabidopsis. Plant J 57:400–412

Revathi S, Subhashree V (2015) Physiological and biochemical mechanisms of heavy metal tolerance. Int J Environ Sci 3:1339–1354

Richau KH, Schat H (2009) Intraspecific variation of nickeland zinc accumulation and tolerance in the hyperaccumulatoro *Thlaspi caerulescens*. Plant Soil 314:253–262

Richau KH, Kozhevnikova AD, Seregin IV, Vooijs R, Koevoets PLM, Smith JAC et al (2009) Chelation by histidine inhibits the vacuolar sequestration of nickel in roots of the hyperaccumulator Thlaspi caerulescens. New Phytol 183: 106–116

Robinson NJ, Winge DR (2010) Copper metallochaperones. Annu Rev Biochem 79:537–562

Saifullah ME, Qadir M, de Caritat P, Tack FMG, Laing GD, Zia MH (2009) EDTA-assisted Pb phytoextraction. Chemosphere 74:1279–1291

Salt DE, Kramer V (2000) Mechanism of metal hyperaccumulation in plants. In: Raskin I, Ensley BD (eds) Phytoremediation of toxic metals. Using plants to clean up the environment. Wiley, New York, pp 231–246

Sanders D, Pelloux J, Brownlee C, Harper JF (2002) Calcium at the crossroads of signaling. Plant Cell Online 14:401–417

Schaaf G, Ludewig U, Erenoglu BE, Mori S, Kitahara T, von Wirén N (2004) ZmYS1 functions as a proton-coupled symporter for phytosiderophore- and nicotianamine-chelated metals. J Biol Chem 279:9091–9096

Socha AL, Guerinot ML (2014) Mn-euvering manganese: the role of transporter gene family members in manga-

nese uptake and mobilization in plants. Front Plant Sci 5:106
Song WY, Park J, Mendoza-Cozatl DG, Suter-Grotemeyer M, Shim D, Hortensteiner S, Geisler M, Weder B, Rea PA, Rentsch D, Schroeder JI, Lee Y, Martinoia E (2010) Arsenic tolerance in Arabidopsis is mediated by two ABCC-type phytochelatin transporters. Proc Natl Acad Sci USA 107:21187–21192
Song WY, Yamaki T, Yamaji KD, Jung KH, Fujii-Kashino M et al (2014) A rice ABC transporter, OsABCC1, reduces arsenic accumulation in the grain. Proc Natl Acad Sci U S A 111:15699–15704. doi:10.1073/pnas.1414968111
Sun H, Xu G, Zhan H, Chen H, Sun Z, Tian B, Hua Y (2010) Identification and evaluation of the role of the manganese efflux protein in Deinococcus radiodurans. BMC Microbiol 10:319
Suresh B, Ravishankar GA (2004) Phytoremediation-A novel and promising approach for environmental clean-up. Crit Rev Biotechnol 24(2–3):97–124, ISSN: 1549–7801
Takahashi R, Ishimaru Y, Senoura T, Shimo H, Ishikawa S, Arao T, Nakanishi H, Nishizawa NK (2011) The OsNRAMP1 iron transporter is involved in Cd accumulation in rice. J Exp Bot 62:4843–4850
Tangahu BV, Abdullah SRS, Basri H, Idris M, Anuar N, Mukhlisin M (2011) A review on heavymetals (As, Pb, and Hg) uptake by plants through phytoremediation. Int J Chem Eng 21:1–31
Thomas K, Anthony RD (2011) Plastic limits of agricultural soils as functions of soil texture and organic matter content. Soil Res 50(1):7–17
Thomine S, Wang R, Ward JM, Crawford NM, Schroeder JI (2000) Cadmium and iron transport by members of a plant metal transporters family in Arabidopsis with homology to Nramp genes. Proc Natl Acad Sci U S A 97:4991–4996
Thomine S, Lelievre F, Debarbieux E, Schroeder JI, Barbier-Brygoo H (2003) AtNRAMP3, a multispecific vacuolar metal transporter involved in plant responses to iron deficiency. Plant J 34:685–695
Verbruggen N, Hermans C, Schat H (2009) Molecular mechanisms of metal hyperaccumulation in plants. New Phytol 181(4):759–776
Verret F, Gravot A, Auroy P, Leonhardt N, David P, Nussaume L, Vavasseur A, Richaud P (2004) Overexpression of AtHMA4 enhances root-to-shoot translocation of zinc and cadmium and plant metal tolerance. FEBS Lett 576:306–312
Wang X, Winter D, Ashrafi G, Schlehe J, Wong YL, Selkoe D, Rice S, Steen J, LaVoie MJ, Schwarz TL (2011) PINK1 and Parkin target Miro for phosphorylation and degradation to arrest mitochondrial motility. Cell 147(4):893–906
Williams LE, Pittman JK, Hall JL (2000) Emerging mechanisms for heavy metal transport in plants. Biochim Biophys Acta 77803:1–23
Xiang C, Werner BL, Christensen EM, Oliver DJ (2001) The biological functions of glutathione revisited in Arabidopsis transgenic plants with altered glutathione levels. Plant Physiol 126:564–574
Xue TT, Li XZ, Zhu W, Wu CG, Yang GG, Zheng CC (2009) Cotton metallothionein GhMT3a, a reactive oxygen species scavenger, increased tolerance against abiotic stress in transgenic tobacco and yeast. J Exp Bot 60:339–349
Yang Z, Chu C (2011) Towards understanding plant response to heavy metal stress. In: Shanker A (ed) Abiotic stress in plants – mechanisms and adaptations. InTech. doi:10.5772/24204
Yang XE, Long XX, Ye HH, He ZL, Calvert DV, Stoffella PJ (2004) Cadmium tolerance and hyperaccumulation in a new Zn-hyperaccumulating plant species (Sedum alfredii Hance). Plant and Soil 259(1–2):181–189
Yang XE, Li TQ, Yang JC, He ZL, Lu LL, Meng FH (2006) Zinc compartmentation in root, transport into xylem, and absorption into leaf cells in the hyperaccumulating species of *Sedum alfredii* Hance. Planta 224(1):185–195
Zacchini M, Pietrini F, Mugnozza SG, Iori V, Pietrosanti L, Massacci A (2009) Metal tolerance, accumulation and translocation in poplar and willow clones treated with cadmium in hydroponics. Water Air Soil Pollut 197:23–34
Zhao K, Liu X, Xu J, Selim HM (2010) Heavy metal contaminations in a soil–rice system: identification of spatial dependence in relation to soil properties of paddy fields. J Hazard Mater 181:778–787. doi:10.1016/j.jhazmat.2010.05.081
Zhao S, Lian F, Duo L (2011) EDTA-assisted phytoextraction of heavy metals by turfgrass frm municipal solid waste compost using permeable barriers and associated potential leaching risk. Bioresource Technol 102:621–626

Metal Tolerance Strategy in Plants

2

Contents

Abstract

The influence of heavy metals on metabolism can be best deciphered through in-depth study of mineral nutrients and heavy metal interaction. Regulatory mechanism pertaining to heavy metal nutrition in plants is dynamic but very scarce literature available. Plants have developed several detoxification mechanisms in order to minimize the detrimental effects of heavy or toxic metal exposure and their accumulation. All these mechanisms are mostly based on chelation and subcellular compartmentalization, and the former is considered as the most common detoxification strategy in different plant species. The main class of metal chelator known in plants is phytochelatins (PCs) and metallothioneins (MTs) which are generally cysteine-rich peptides, besides, normal expression of GSH/PCs during metal exposure, certain genetic manipulations of these substances help plants to improve toxicity of heavy metals. The cytosol is considered as an important metal detoxification mechanism under stress conditions. The metal chelation by PCs in the cytosol of plants is known to be the best heavy metal detoxification mechanism and tolerance.

Keywords

Chelation • Cytosol • Metallothioneins • Phytochelatins • Toxic metal

S. Jan, J.A. Parray, *Approaches to Heavy Metal Tolerance in Plants*,
DOI 10.1007/978-981-10-1693-6_2

2.1 Heavy Metal Interaction with Other Nutrients

The influence of heavy metals on metabolism can be best deciphered through in-depth study of mineral nutrients and heavy metal interaction. Regulatory mechanism pertaining to heavy metal nutrition in plants is dynamic but very scarce literature available. Plant response to a combination of metals in the growth media has been varied; it had been classified to following three basic groups according to Nagajyoti et al. (2010).

Plant exhibits preservative response in terms of plant growth assessment under cumulative metal stress, and these responses were similar in which investigated metals were supplied discretely. Some other plants demonstrate antagonistic responses in which plant growth parameters under cumulative metal stress were relatively immense than in plants in which heavy metals were added independently. Iron and zinc are antigonistic which explains depression in growth of corn plants supplied with low and high iron levels (Masarovicova et al. 2011). Zn accumulation affects iron translocation more than its absorption resulting in iron deficiency due to Fe–Zn competition (Masarovicova et al. 2010). Other groups of plants display a synergistic approach in which the response of plant development parameters in collective heavy metal stress was diminutive than in plants supplied with metals autonomously. In numerous plants, copper seemed to exhibit synergistic response to other metals like zinc and nickel and antagonistic to cadmium and additive with cobalt (Pesko 2009). Heavy metals are known to interfere with mobility and availability of iron uptake resulting in iron deficiency. Heavy metal accumulation has been found to be detrimental for copper and zinc uptake as well. The susceptibility order of these trace metals to heavy metals seems to be Fe >>Zn ~ Cu. Heavy metal accumulation also resulted in disturbances of Ca/K ratio leading to tremendous hurdles in water transport. Heavy metal accumulation results in interruption of basic plant nutrient uptake (Zuo and Zhang 2011). Oat plants treated with Ni, Co, and Zn had even three times higher phosphorus levels in shoots than control plants. Iron seems to be the important targets of toxic metal influence on metabolism of plants. There are ample studies confirming heavy metal accumulation like Cd, Co, Cr, Pb, Mn, and Zn causing iron deficiency (Dixit et al. 2015). Graminaceous crop species exhibit intense Zn and Fe interaction owing to the presence of phytosiderophores which enhance zinc absorption (Rojas et al. 2011). Iron is also known to significantly decline Mn uptake in plant supplemented with medium having high iron content or above optimum level (Madejczyk and Ballatori 2012). Copper and iron toxicity in rice plant inhabiting water logging field was subsided by providing iron supply. Bioavailability of some toxic metals like Cd was enhanced by root exudates (Mench and Martin 1991). Proper iron supply was found to significantly diminish cadmium content in lettuce leaves with parallel increment in K and Zn content (Zuo and Zhang 2011), and cadmium in turn was known to interfere with uptake of other nutrients through its influence on membrane permeability (Samaranayke et al. 2012). Cadmium uptake was decreased by its interaction with Pb and Zn, while their independent intake was increased radically (Bernhoft 2012). Pb uptake was also decreased by Zn, but Pb itself resulted in Ca and Mn depletion in *Picea abies* needles via decreased apoplastic sap flux through endodermis (Yin et al. 2010). However, increment in essential metallic nutrients may be precarious for HM-treated plants. Increment in Co, Cr, Zn, Mn, and Mo content intensifies the nickel-induced chlorosis and necrosis. Mn and Mg interaction in terms of Mn/Mg ratio is a significant factor for maintaining endurance of plants even if growing under phytotoxic Mn concentration (Troadec et al. 2010). Nitrogen and copper had an antigonistic influence on their concentration in plants; Weber et al. (1991) noticed significant decline in nitrogen in the form of NO_3^-/NO_2^-, NH_4^+, amino acid-N, and protein-N in copper-treated *Silene vulgaris* (Moench.) shoots with no change in nitrogen levels of roots (Moulick et al. 2013) (Fig. 1).

2.2 Inversion of Metal Toxicity with Nutrient Element Interactions

Numerous experiments were performed on reversing metal toxicity by elevated macronutrient supply. Most ancient work performed by Bernhoft (2012) showed increased K and P supply caused significant decline in Cd absorption. Plants with relatively high K supply accumulated less Cd and Pb than the plants with comparatively lower potassium levels (Shamsi et al. 2008). Reversal of heavy metal toxicity by ameliorating K and P levels could be explained by a competition between these elements (Flora and Pachauri 2010). Heavy metal is used for detoxification by intensifying the uptake and transport of other nutrients or by inhibiting the uptake of other heavy metals. Competition in uptake was strong among elements with similar chemical properties like Zn, Pb, and Cd (Troadec et al. 2010). Translocation of essential nutrients, i.e., Mn, Mg, Cu, etc. from the roots to aboveground tissues in Cd-treated plants was stimulated by augmenting Mo levels resulting in stimulatory effect on plant metabolism. Increased Mn or Fe supply also decreased Cd levels in leaves (Klaassen 2006). Similarly, nickel-induced chlorosis was modulated by supplementing Ni-treated plants with higher iron and nitrogen supply. Application of macronutrients like P or metallic elements like K or Fe may revitalize HM-treated plants. However, in vivo studies cannot be transferred to natural conditions because of highly complicated relation among all nutrients and varied plant susceptibility at particular growth stages (Naser et al. 2011). Physiological assessment of plants susceptible to HMs also depends on primary target organs. Toxic metal ions like, Cd, induce nutrient deficiency by decreasing the concentration of some essential nutrients in plants (Guo et al. 2004), and it is inevitable to mitigate the metal-induced detrimental effects by providing mineral nutrition to plants. Plant nutrients are generally known to influence the action and bioavailability of heavy metals like Cd in the plant–soil sphere. Plant nutrients show both direct and indirect effects on Cd mobility and its toxicity. The indirect effects include increasing the Cd levels by elevating plant biomass and physiological stress (Sarwar et al. 2010). The direct impacts are more pronounced in terms of Cd mobility in soil by encouraging precipitation or adsorption, competition, and also Cd sequestration particularly in vegetative parts rather than in other edible parts. For example, phosphate ions promote the Cd^{2+} precipitation (Hong et al. 2008), while Fe^{2+} competes with Cd^{2+} for the same membrane transporters in plant cells (Vert et al. 2002; Kovacs et al. 2009). Xie et al. (2009) have observed that phytoextraction of Cd and Zn in *Thlaspi caerulescens* was enhanced by nitrate (NO_3^-) supply. Previously, the effect of N deficiency, or application of either nitrate or ammonium on plants and subsequent defense responses have mostly been studied (Engelsberger and Schulze 2012). There are few reports pertaining to impact of excessive plant nutrition on induced defenses in plants (Lin et al. 2012). Further, the surplus use of N fertilizers in agriculture can become a pollutant (Savci 2012) and also for the increasing the protein content in different crops led to the elevated Cd concentration in plants grown on contaminated soils (Hasan et al. 2009). Nitrogen is considered as important regulating factor for heavy metal mobility and availability for plants by soil acidification. However, excessive N is reported to inhibit the metabolism of plant N compounds that perform signal functions (Nagajyoti et al. 2010). For example, Konotop et al. (2012) observed the decreased Cd uptake in vicinity of higher nitrogen *Glycine max*.

There are many thoughts regarding the regulation of genes that increased screening of nitrate in plants roots. For example, the induction of NRT1.8 expression and repression of NRT1.5, specific transporters by Cd^{2+}in *Arabidopsis*, is responsible loading/unloading of nitrate into/from the xylem sap, respectively (Hasan et al. 2009). The Cd and N interactions caused decreased in Cd uptake by soybean seedling roots under high nitrogen nutrition level, and plant nitrogen fertilization can be used on cadmium-contaminated soil (Konotop et al. 2012). The accumulation of nitrate in the plant roots serves as a signal to enhance nitrate assimilation like expression of some N containing compounds like glutathione and phytochelatins for detoxification of Cd^{2+} (Gao et al. 2011). It is vital to mitigate some of the metal-induced negative effects by optimization of plant mineral nutrition as toxic metals like, Cd, induce nutrient deficiency or even decrease macronutrient concentrations in plants (Gao et al. 2011). The plant nutrients are generally known to affect the activity and bioavailability of Cd in the soil–plant environment. The indirect effects include further dilution of Cd ion concentration by elevated plant biomass and accordingly physiological stress (Leitenmaier and Kupper 2011) while as direct impacts causes decreased Cd solubility in soil via precipitation, adsorption, competition between metal ion and nutrients for the same membrane transporters, and Cd sequestration in vegetative parts avoiding its accumulation in edible parts of the plants (Xie et al. 2009).

N is an essential macronutrient and is an important component of many structural, genetic, and metabolic compounds in plants (Hussain et al. 2009; Konotop et al. 2012).). The two plant hormones are known to be transported auxin and abscisic acid (ABA). The abscisic acid is a ubiquitous phytohormone involved in many developmental processes and stress responses of plants (Gao et al. 2011). The identification of the ATP-binding cassette (ABC) transporter in *Arabidopsis thaliana* and pleiotropic drug resistance transporter PDR12 (AtPDR12)/ABCG40 as a plasma membrane AB uptake transporter have been documented (Liu et al. 2011; Konotop et al. 2012).

Polyamines (PAs) and brassinosteroids (BRs) are well-known growth regulators playing important roles in stress management among plants (Hasan et al. 2008). PAs are small aliphatic nitrogenous compounds with ubiquitous distribution. The reduction of Cr stress by EBL or Spd was also supported by enhanced values of stress indices like PCs, photosynthetic pigments, total soluble sugars, and reduction in malondialdehyde and peroxide levels in Cr-treated seedlings (Bhardwaj et al. 2007).

Insinuation of PAs in improving of various abiotic and biotic stresses has made them an essential component of plant defense and detoxification mechanisms. The co-application of BRs and PAs is more significant in improved Cr stress than individual treatments (Hasan et al. 2008) which involve mitigation with EBL and Spd co-application that led to the physiological and molecular interactions in a additive manner (Konotop et al. 2012). Hence, providing a sole and eco-friendly strategy employing interplay of BRs and PAs to overcome heavy metal stress mitigation, and abiotic stress like in radish (Choudhary et al. 2012). Several reports have shown the overexpression of different genes encoding LOXs under heavy metal stress (Remans et al. 2010). Similar observations were made by Maksymiec et al. (2005) in *Arabidopsis* under Cd or Cu stress. Mostofa et al. (2015) have also showed that Cu induced a rapid increase of jasmonic acid (JA) content in *Oryza sativa* leaves, and it may be the reason of the usually observed rapid decrease in growth processes and senescence intensification of plants after long heavy metal exposure. Ethylene is also reported to be a part of the signalling process under metal exposure (Arteca and Arteca 2007). It is mentioned that ethylene production was stimulated under Cu stress by an augmentation of ACC synthase enzymatic activity in *Arabidopsis* plants (Arteca and Arteca 2007). Ethylene acts by the activation of the mitogen-activated protein kinase (MAPK) cascade, a signal transduction process which is activated under Cd and Cu (Schellingen et al. 2014). It is also reported that the interactions between different phytohormones, like JA, ethylene, and other ones expediate the gene expression of cytochrome P450 under both biotic and abiotic stresses (Mostofa et al. 2015). Metwally et al. (2003) however reported that Cd toxicity might be increased by (SA) salicylic acid (signalling factor), which often blocks the JA pathway (Takahashi et al. 2007). Secondary metabolites and PAs have been mentioned in many heavy metal-stressed plants where transcription is mediated by JA or ethylene suggesting that biotic and abiotic stresses may share some signalling processes (Schellingen et al. 2014). Further, Masood et al. (2012) observed that the preexposure to a mild dose of Cd ions increased plant resistance to both viral and fungus infection, and Hossain et al. (2012) reported that plants may increase tolerance to pathogens by induction of defenses like glucanases, chitinases, or proteinases under heavy metal stress.

2.3 Role of Phytochelatins in Metal Tolerance

Plants have developed several detoxification mechanisms in order to minimize the detrimental effects of heavy or toxic metal exposure and their accumulation. All these mechanisms are mostly based on chelation and subcellular compartmentalization, and the former is considered as the most common detoxification strategy in different plant species. The main class of metal chelator known in plants is phytochelatins (PCs) which are generally cysteine-rich peptides (Blindauer 2008). The PCs are mainly synthesized from reduced glutathione (GSH) in a transpeptidation reaction catalyzed by the enzyme phytochelatin synthase (Hassinen et al. 2009). So GSH is considered as a limiting factor for PC synthesis in plants particularly under heavy metal stress. Besides normal expression of GSH/PCs during metal exposure, certain genetic manipulations of these substances help plants to improve toxicity of heavy metals (Vinita 2007). As mentioned, PCs are low-molecular-weight metal-binding polypeptides with the structure as (γ-Glu-Cys) nGly (n = 2–11). The PCs are present all living organisms including fungi, plants, etc. (Cobbett and Goldsbrough 2002). The PCs form complexes with toxic metal ions in the cytosol and are then transported into the vacuole for plant protection by metal toxic effects (Hassinen et al. 2009). The PC synthesis in response to Pb or Cd and formation of PC–Pb/PC–Cd complex is also well documented (Marentes and Rauser 2007). There are different deliberations on PC translocation as mentioned in *Arabidopsis* that they undergo long-distance transport between roots and shoots. Similarly, translocation of PCs has been identified using xylem and phloem sap from *Brasica napus* (Cobbett and Goldsbrough 2002). Further, the high PCs/Cd and GSH/Cd ratios in the phloem sap suggested that PCs and GSH can perform as long transporters of Cd (Anjum et al. 2012) and only traces of PCs were seen in xylem sap. Consequently, the phloem seems to be a major vascular system for long-distance source to sink transport of cadmium which is transported as PC–Cd and GSH–Cd complexes (Miguel et al. 2013).

2.4 Metal Complex Formation by PCs

Plants generally produce two metal-binding ligands, i.e., PCs and metallothioneins (MTs) (Yadav 2010). The cytosol is considered as an important metal detoxification mechanism under stress conditions. Molecular genetic approaches have helped to understand the advances in biosynthesis and function of PCs (Cobbett and Goldsbrough 2002). PCs are a family of cysteine-rich polypeptides with the general structure (γ-Glu-Cys)*n*-X, in which X is Gly, γ-Ala, Ser, Gln, or Glu and *n* = 2–11 depending on the organism, although the most common forms have 2–4 peptides (Clemens and Persoh 2009). PCs are synthesized from GSH; the metal binds to the constitutively expressed enzyme γ-glutamyl cysteinyl dipeptidyl transpeptidase (PC synthase), thereby activating it to catalyze the conversion of GSH to phytochelatin (Yadav 2010). The induction and biosynthesis of PCs mainly occur under heavy metal exposure like Cd, Hg, Ag, Cu, Ni, Au, Pb, As, and Zn; among all, the Cd is regarded as the strongest inducer of PCs. Generally, PCs chelate Cd^{+2} ions through the –SH of cysteine and are accumulated in the vacuole by ABC transporters, accordingly limiting the exchange of free Cd^{2+} inside the cytosol (Zhao et al. 2010). Plants always adopt the strategy of making Cd–GSH and Cd–PCs complexes to sequester Cd within vacuoles efficiently and also transport Cd through both xylem and phloem vessels (Yadav 2010). The PC induction under metal exposure was presented by the fact that their induction is positively correlated with metal accumulation in all plant tissues (Clemens and Persoh 2009). For example, under Cd stress, Cd^{2+} was taken up by *Eichhornia crassipes* (water hyacinth) roots and was present as Cd-binding complexes (Noctor et al. 2012). Further, the characterization of Cd-binding complex was observed to have two cadystins A, (γ-Glu-Cys)3-Gly, two cadystins B, (γ-Glu-Cys)2-Gly, and inorganic sulfur (Clemens and Persoh 2009). In another study, Iglesia-Turino et al. (2006) reported mercury accumulation in *Brassica napus* under high Hg mercury concentration, and Hg accumulation was strongly influenced with PC concentration. PCs play an important role in the $Zn^{2+}/Cu^{+}/Cu^{2+}$ homeostasis by allowing brief ion storage (Noctor et al. 2012). The role of PCs in the metal detoxification vis-à-vis mitigation of oxidative stress and phytochelatin synthase gene expression has been reported in

Allium sativum L. and *Dunaliella tertiolecta* (Verret and Gravot 2004). Although sufficient data is available regarding the role of PCs in HM detoxification and accumulation in higher plants and complex formations (Cobbett and Goldsbrough 2002), however, the more study of role of PCs in the metal tolerance mechanism at the cellular level is the need of hour.

2.4.1 Metal Chelation with Reference to Cadmium by Phytochelatins (PCs)

The metal chelation by PCs in the cytosol of plants is known to be the best heavy metal detoxification mechanism and tolerance (Ali et al. 2013). Metallothionein and phytochelatin are low-molecular-weight peptides with a high specific metal-binding capacity. Earlier all metal-binding proteins in plants were assumed to be MTs, and in an old classification system, phytochelatins were described as enzymatically synthesized class III (Nishiyama et al. 2013). However, it is apparent now that plants express both PCs and MTs and they independently take part in metal detoxification or metabolism (Cobbett and Goldsbrough 2002).

PCs have been found in many plants and other photosynthetic organisms like algae, gymnosperms, and from monocots to dicots. Phytochelatins are generally synthesized from GSH by enzymes, i.e., PC synthases. These PCs help in distribution and accumulation of highly toxic metals like Cd, Ag, Hg, Cu, etc. (Chu et al. 2010).

However, in this section, we will mainly focus on the role of PCS in phytodetoxification of Cd ions. There are numerous techniques like X-ray absorption spectroscopy, HPLC-MS, and inductively coupled plasma optical emission spectrometer (ICPOES) to reveal that Cd^{2+} is bound to PCs in plants. For example, phytochelatins bound 82 %, 19 %, and 12 % of the Cd in roots and in young and old leaves in wheat, respectively (Marentes and Rauser 2007). In one study by Nagasaka et al. (2009). Cd bound PC percentage in Indian mustard seedlings increased from 34 % to 60 % from 6 h of Cd exposure to 72 h. The PCs have also been reported in Cd-hyperaccumulator desert plants *Salsola kali*, vis-à-vis Cd storage in cells (Cobbett and Goldsbrough 2002). Similarly, PC–Cd complexes under Cd exposure have been seen in mushrooms like *Boletus edulis* with reduced level of GSH (Bailey et al. 2012).

The chemical structure of thiol and carboxyl groups is considered to be vital for the metal-binding capability and formation of Cd–PC complexes (Satofuka et al. 2001). Two types, i.e., low-molecular-weight and high-molecular-weight Cd–PC complexes, are formed during Cd sequestration where in LMW complexes serves as the temporary form for transporting Cd^{2+} from cytosol to vacuole, and in later more, Cd and sulfide ions are included to produce the HMW complexes (Miller et al. 2008). The *S. pombe hmt1* mutant seems to be first in transporting Cd^{2+} via PC–Cd complex. The SpHMT1 (ATP-binding cassette (ABC) transporter protein) is situated in the vacuolar membrane and arbitrates the ATP-required transport of LMW PC–Cd complexes into vesicles (Navarro et al. 2009). Some ATP-binding cassette (ABC) transporters in *Arabidopsis* genome determined by cDNA-microarray approach were observed to be differentially regulated under Cd stress suggesting their role in Cd sequestration and redistribution (Bovet et al. 2005). Similarly, MRPs (multidrug resistance-associated proteins), the subfamily of ABC transporters, are involved in PC–Cd complex transport across the tonoplast in some plants (Rea 2007). In addition to Cd–PC complexes, more specific PC vacuolar transporters (ABCC type) for other heavy metals have been identified in *Arabidopsis*, i.e., AtABCC1 and AtABCC2 mediating transport of As(III)-PC (Korenkov et al. 2007). In some cases, cation/proton exchanger (CAX) transporters transport and distribute Cd^{+2} through vacuole. For example, the pH-dependent Cd^{2+} accumulation in vesicles was accompanied by efflux of protons in oat roots (Guo et al. 2004). The analysis of CAX2 transporters among various plant species like *Arabidopsis* and tomato barley reveals that there are diverse and specific regulatory mechanisms with regard to CAX transporters (Bailey et al. 2012).

2.5 Chelation of Heavy Metals by Metallothioneins (MTs)

MTs have a low molecular weight of 4–8 kDa, and are cysteine-rich polypeptides that have a tendency to bind toxic metals by the SH group of cysteine (Hazrat et al. 2013). The main role of MTs includes the maintenance of homeostasis of essential metals, toxic metal sequestration, and protective function against oxidative cell damage (Clemens and Persoh 2009). Due to their diverse features, MTs have been subdivided into three classes based on Cys residue arrangements (Bell and Vallee 2009) which include Cys-(X)-Cys, Cys-Cys, and Cys-(X-X)-Cys motifs where X denotes amino acid. The architect or allocation of cyst residues determines various MT isoforms and accordingly their ability to bind and detoxify toxic metals (Cobbett and Goldsbrough 2002). The biosynthesis of MTs is synchronized at the transcriptional level which is induced by several factors, including hormones and cytotoxic agents, and by toxic metals like Cd, Zn, Hg, Cu, Au, Ag, Co, Ni, etc. (Xue et al. 2009). Further, there are numerous studies related to gene expression related to their induction at different developmental stages and under toxic metal stress conditions, and these MT genes emerge to behave differentially in a tissue-specific manner under toxic metal stress (Ngu et al. 2010). In *Brassica rapa*, three MT genes (*BrMT1*, *BrMT2*, and *BrMT3*) regulate differently under toxic metal stress conditions. *BrMT1* and *BrMT2* were not significantly induced, while elevated expression of *BrMT3* was apparent in Fe-treated seedlings (Ahn et al. 2012). *BrMT1* expression was elevated in Cu-treated seedlings after 3 h treatment, while *BrMT2* expression was downregulated and *BrMT3* remained unchanged (Barry et al. 2011). Likewise, *BrMT1* increased a little, while *BrMT2* was downregulated and *BrMT3* remained unchanged in Zn-treated seedlings, while in Mn-treated seedlings, *BrMT1* and *BrMT3* genes were shown increase expression up to 12 h and after that were downregulated (Park et al. 2012). Various data documented the role of MTs in metal detoxification and homeostasis, but metal-inducibility of plant MTs has not been yet reported (Ahn et al. 2012). Further, the selection of model system as well as model hyperaccumulator like *Arabidopsis halleri* and some other species like *Thlaspi* spp. is likely to further elucidate the molecular mechanisms of metal transport, tolerance, and homeostasis in plants (Castiglione et al. 2007). Further, information regarding the structures and properties of MTs could clarify their mechanisms of action and functions.

2.6 Metal Detoxification by Organic Acids, Amino Acids, and Other Phosphate Derivatives

Besides PCs and MTs, toxic metal tolerance and detoxification in plants by chelation can be achieved by secretion of acids and phosphate derivatives. Organic acids are excreted from plant roots and play a role in the external heavy metal detoxification process by forming stable metal ligand complexes. These complexes transform their bioavailability and mobility of metal ions and restrict their entry into plants or inhibit their accumulation in roots. And in the cytosol, internal metal detoxification by organic acids can be achieved by chelation with heavy metals and is to be transformed into nontoxic or less toxic substances (Hall 2002). Plants produce different types of ligands like carboxylic acid (citrate, malate, and oxalate) and amino acids (histidine and nicotianamine), and phosphate derivatives (phytate) are potential ligands for chelation of heavy metals like Al, Cd, Cu, Ni, Co, and Zn and for tolerance and detoxification (Cobbett and Goldsbrough 2002). Citrate is known to have high affinity for chelation of toxic metals like Fe, Cd, Ni, Co, and Zn (Llamas et al. 2008). Several Ni hyperaccumaltors has higher affinity with the citrtate and the amount of citrate produced was found directly proportional to accumalated Ni in plant cells (Kramer et al. 2000). After that, many reports mention about the role of carboxylic acids for metal detoxification.

Further, the citric acid has been reported to be a major ligand at low Cd concentrations (Ma et al. 2005) and also found to contribute in Zn tolerance and accumulation (Weber et al. 2004). Similarly, the correlation between citric acid exudation and aluminum tolerance was seen in *Phaseolus vulgaris* (Wang et al. 2004). The co-hyperaccumulator, i.e., *Crotalaria cobalticola*, and some non-accumulators *Rauvolfia serpentina* and *Silene cucubalus* to Co ions increase the citrate concentration and vis-à-vis involve in chelation and metal detoxification (Weber et al. 2004). Besides other root secretion like oxalate and then Al detoxification have also been documented (Ma et al. 2005). In one study, oxalic acid from the roots in response to Al stress and accumulation of nontoxic Al oxalate in the leaves of *Fagopyrum esculentum* Moench for the external and internal detoxification has been reported (Weber et al. 2004). Malate was secreted in cytosol for Zn chelation in Zn-tolerant plants (Clemens and Persoh 2009). It is mentioned that there are differences within the plant species for the Al tolerance and on the malate transport out of the apical root cells via an Al-activated malate permeable channel (Zhao et al. 2009). The Al-tolerant genotypes were reported to secret about tenfold higher malate and fivefold higher succinate than Al-sensitive seedlings over 24 h exposure to 50 μM malate. Vander et al. (2013) observed that some amino acids like histidine also play a significant role in the metal chelation within plant cells and in the xylem sap (Gill et al. 2014). For example, Ying et al. (2010) reported Zn–His complexes in the *Thlaspi caerulescens*. Further, free histidine in the xylem exudates of a Ni hyperaccumulator like *Alyssum lesbiacum* was increased up to 36-fold under Ni stress. The transgenic plants have also been documented for metal detoxification like overexpressing the *Salmonella typhimurium* ATP phosphoribosyl transferase gene (*StHisG*) that shows twofold higher histidine levels than wild plants (Lamas et al. 2008). Nonproteinogenic amino acids (NA) are free amino acids that are observed in root and leaf cells as well as in phloem sap and are mainly synthesized by a one-step condensation reaction of three (03) molecules of S-adenosyl-Met (SAM) by NA synthase (Rellan-Alvarez et al. 2008). NA are mainly responsible for binding of toxic metals like Fe, Zn, Al, Cu, and Ni (Kawachi et al. 2008). NA are reported to help in Fe homeostasis besides their role in long-distance metal transport in solanaceous plants (Brunetti et al. 2011). For example, *Arabidopsis* gene chips were used to identify those genes that are more active in roots of *A. halleri* than *A. thaliana* under controlled conditions, and two genes showing highest levels of expression in *A. halleri* roots code for an NA synthase and a putative Zn^{2+} uptake system (Weber et al. (2004).

Phytates have also been reported for heavy metal complexing, and they tend to be associated with metal deposits in plants like Al-tolerant *Zea mays* vacuoles (Bose et al. 2010). The Zinc–phytate globules are commonly seen in tolerant ecotypes of *Deschampsia cespitose* roots than in sensitive ecotypes (Ying et al. 2010; Cobbett and Goldsbrough 2002). In this regard, more research is required to evaluate the organic acids for toxicity versus tolerance response over a range of plant species. However, it is known that plant adaptation to metal stress is controlled by genetically determined signalling system and is species specific (Ma et al. 2005). Therefore, the identification and characterization of plant genes which control responses to metal stresses are a vital move to evaluate the complex regulatory network for determination of metal tolerance in plants (Yu et al. 2003). The constitutive gene expression determines the molecular adoption of plants which might be essential for genetic transformation and phytoremediation (Guo et al. 2008; Hossain et al. 2012). For example, the expression of glutathione (GSH) metabolic genes was studied in Cd and Hg-treated alfalfa seedlings, as well as the GSH level, homoglutathione (hGSH), and AsA. The expression of genes, i.e., *ECS*, *GS*, *hGS*, *GR1*, *GR2*, *GPX*, and *PCS*, involved in their metabolism was analyzed (Yang et al. 2014). 30 μM Cd for variable exposure periods did not alter the expression of the genes; however, 30 μM Hg was seen to affect the accumulation of some of the transcripts. Hg produced a clear induction of *GR1* (cytosolic) and *GR2* (plastidic) genes and with a maximum expression at 3 h after metal exposure. Similarly, *ECS* and *hGS* genes were expressed slowly and also reaching maximum transcript accumulation after 3 h (Wang et al. 2012).

The induction of the GSH metabolic gene in response to Cd stress was reported in *Camellia sinensis*, and transcript levels of GSH biosynthetic and regenerating genes, i.e., γ-ECS, GS, and GR, were increased under Cd stress (Rausch et al. 2007). Further, the GST transcript levels did not show any alter following Cd treatment (Mohanpuria et al. 2007) and also upregulated the GSH level of different metabolic genes suggesting the role of these enzymes in the protection of plants from a high level of Cd exposure (Ortega-Villasante et al. 2005).

Li et al. (2004) have also shown that rice roots under Cd stress increased GST (GSTs: *Os*GSTU3, *Os*GSTU4, and *Os*GSTU12) and APX gene expression. The induction of the OsGST gene (*Os*GSTU5) in response to Cd stress is involved in the direct quenching of Cd ions forming GSH–Cd complexes (Villasante et al. 2007; Guo et al. 2008). Under Al stress, GST and CAT gene expression was found higher in roots of *Pisum sativum* L., whereas CAPX increased in shoots (Benavente et al. 2004; Sato et al. 2011).

The Cd stress induced upregulated expression of Fe-SOD, Mn-SOD, Chl Cu/Zn-SOD, Cyt Cu/Zn-SOD, APX, GPX, GR, and POD—and it was concluded that genes involved in antioxidant enzymes could play important roles in Cd tolerance in perennial ryegrass (Benavides et al. 2005; Jia et al. 2010; Hossain et al. 2012). The transcript abundance of various genes involved in antioxidative defense in root and leaf tissues was observed. The generation of oxygen radicals is the main cause of oxidative stress in the roots, and peroxide appears to play an important role in leaves (Cho and Seo 2005). Further, an increased transcript level of LOX (a potential inducer of oxidative stress) proposed a key role in causing the Cd-related redox imbalance (Xiang et al. 2001), and this led to the fact that the antioxidative defense system is moderated by the activation of various genes in different organs and cellular compartments (Guo et al. 2008).

Goupil et al. (2009) studied the identification of various genes involved in metal tolerance *Lycopersicon esculentum* cv. VFN-8 plants. They were subjected to As(V) or Cr(VI) stress, and transcript analysis of genes using RNA isolated from tomato roots or shoots treated for 24 h with As(V) or Cr(VI) at concentrations ranging from 80 to 640 μM was studied. It was observed that an upregulation of *Hsp90-1* transcript was due to both of the metal stress treatments (Hao et al. 2011). Further, the tomatoes showed phenotypic symptoms of metal toxicity Cr(VI) stress but not to As(V), representing that tomatoes are more tolerant to As than Cr (Goupil et al. 2009). The relative contribution of PCs and MTs in toxic metal tolerance, i.e., Cd or Cu, was studied in black mangrove, i.e., *Avicennia germinans*, and gene expression was evaluated, and it was observed that a low concentration of Cd and Cu led to a significant increase in *AvPCS* expression after 4 h exposure, while a sharp increase in *AvMt2* expression was observed after 16 h metal exposure (Cobbett and Goldsbrough 2002; Hassinen et al. 2011).

Duressa et al. (2011) conducted a detailed transcriptomic study to determine the genetic factors for the mechanism of Al tolerance in Al-tolerant soybean line (PI 416937), and they revealed that GST and GR transcript levels increase in response to Al stress (10 μM Al) indicating that the ROS metabolic genes had a definite role in Al tolerance in plants Ezaki et al. 2000). Further, the gene expression of CS, GST-omega, and GR was studied, and these genes exhibited very similar expression patterns at both protein and mRNA levels (Cobbett and Goldsbrough 2002). Thus, these findings provided the additional evidence for correlative expression/accumulation of the identified proteins and gene expression in rice plant roots under As stress.

Further, the study on *Phragmites australis* under zinc stress revealed that most Zn was accumulated in the apoplast or sequestered into the vacuolar lumen (Fukao et al. 2011). It was documented that CDF (cation diffusion facilitator) transporters earlier called MTP (metal tolerance protein) are involved in mediating the cytoplasmic efflux of various metal cations, i.e., Zn^{2+}, Cd^{2+}, Co^{2+}, Ni^{2+}, and Mn^{2+}. Ashrafi et al. (2011) also study overexpression lines of two proteins, namely, COT1 and ZRC1 in cobalt and zinc/cadmium tolerance, and it was observed that both are localized to the vacuolar membrane, indicating a role in metal sequestration in *S. cerevisiae*

(Ashrafi et al. 2011). The *Arabidopsis* ZAT/MTP1, a member of CDF family, was first identified for a possible vacuolar Zn transporter in plants. AtMTP1 is generally localized to vacuolar membranes, and the overexpression of protein-coding domain of ZAT results in increased Zn resistance and vis-à-vis accumulation in the roots under high Zn exposure (Kobae et al. 2006). Some vacuolar membrane-localized TgMTP1 transporters of hyperaccumulator *Thlaspi goesingense* show tolerance to a broad spectrum of metal ions like Ni, Cd, Zn, and Co and harmonize the metal sensitivity of the yeast *COT1/ZRC1* mutant strains and surely increase the zinc tolerance by upregulation of Zn transporter genes (*ZIP3*, *ZIP4*, *ZIP5*, and *ZIP9*) (Gustin et al. 2009).

Progressive findings indicated that *Arabidopsis* NRAMP proteins have an pivotal role in Mn homeostasis and Cd toxicity. The *nramp3nramp4* double mutant showed lower Mn release from mesophyll vacuoles, and it's suggested that AtNRAMP3 and AtNRAMP4 export Mn from vacuoles to maintain mitochondrial Mn-SOD activity under Mn deficiency (Lanquar et al. 2010).

References

Ahn YO, Kim SH, Lee J, Kim HR, Lee HS, Kwak SS (2012) Three Brassica rapa metallothionein genes are differentially regulated under various stress conditions. Mol Biol Rep 39:2059–2067

Ali H, Khan E, Sajad MA (2013) Phytoremediation of heavy metals–concepts and applications. Chemosphere 91:869–881

Anjum NA, Ahamd I, Mohmood I, Pacheco M, Duarte AC, Pereira E (2012) Modulation of glutathione and its related enzymes in plants' responses to toxic metals and metalloids–a review. Environ Exp Bot 75:307–324

Arteca RN, Arteca JN (2007) Heavy-metal-induced ethylene production in *Arabidopsis thaliana*. J Plant Physiol 164:1480–1488

Ashrafi E, Alemzadeh A, Ebrahimi M, Ebrahimie E, Dadkhodaei N, Ebrahimi M (2011) Amino acid features of P1B-ATPase heavy metal transporters enabling small numbers of organisms to cope with heavy metal pollution. Bioinf Biol Insights 5:59–82

Bailey SJ, Fukao T, Gibbs DJ, Holdsworth MJ, Lee SC, Licausi F (2012) Making sense of low oxygen sensing. Trends Plant Sci 17:129–138

Barry AN, Otoikhian A, Bhatt S, Shinde U, Tsivkovskii R, Blackburn NJ, Lutsenko S (2011) The lumenal loop Met672-Pro707 of copper-transporting ATPase ATP7A binds metals and facilitates copper release from the intramembrane sites. J Biol Chem 286(30):26585–26594

Bell SG, Vallee BL (2009) The metallothionein/thionein system: an oxidoreductive metabolic zinc link. Chembiochem 10(1):55–62

Benavides M, Gallego MS, Tomaro ML (2005) Cadmium toxicity in plants. Braz J Plant Physiol 17(1):21–34

Bernhoft RA (2012) Mercury toxicity and treatment: a review of the literature. J Environ Public Health 2012:460–508

Bhardwaj R, Arora N, Sharma P, Arora HK (2007) Effects of 28-Homobrassinolide on seedling growth, lipid peroxidation and antioxidative enzyme activities under nickel stress in seedlings of Zea mays L. Asian J Plant Sci 6(5):765–772

Blindauer CA (2008) Metallothioneins with unusual residues: histidines as modulators of zinc affinity and reactivity. J Inorg Biochem 102:507–521

Bose J, Babourina O, Shabala S, Rengel Z (2010) Aluminium-induced ion transport in Arabidopsis: the relationship between Al tolerance and root ion flux. J Exp Bot 61:3163–75

Bovet L, Feller U, Martinoia E (2005) Possible involvement of plant ABC transporters in cadmium detoxification: a cDNA sub-microarray approach. Environ Int 31:263–267

Brunetti P, Zanella L, Proia A, De Paolis A, Falasca G, Altamura MM, Sanita di Toppi L, Costantino P, Cardarelli M (2011) Cadmium tolerance and phytochelatin content of Arabidopsis seedlings overexpressing the phytochelatin synthase gene AtPCS1. J Exp Bot 62:5509–5519

Castiglione S, Franchin C, Fossati T, Lingua G, Torrigiani P, Biondi S (2007) High zinc concentrations reduce rooting capacity and alter metallothionein gene expression in white poplar (Populus alba L. cv. Villafranca). Chemosphere 67:1117–1126

Cho UH, Seo NH (2005) Oxidative stress in Arabidopsis thaliana exposed to cadmium is due to hydrogen peroxide accumulation. Plant Sci 168:113–120

Choudhary S, Ansari MYK, Aslam R (2012) Sequential effects of cadmium on plant growth, biochemical and cytophysiological aspects, antioxidant activity, and molecular screening of protein by SDS–PAGE in *Trigonella*. Toxicol Environ Chem 94(8):1557–1570

Chu HJ, Lin YP, Jang CS, Chang TK (2010) Delineating the hazard zone of multiple soil pollutants by multivariate indicator kriging and conditioned Latin hypercube sampling. Geoderma 158(3–4):242–251

Clemens S, Persoh D (2009) Multi-tasking phytochelatin synthases. Plant Sci 177:266–271

Cobbett C, Goldsbrough P (2002) Phytochelatins and metallothioneins: roles in heavy metal detoxification and homeostasis. Annu Rev Plant Biol 53:159–182

Dixit R, Malaviya D, Pandiyan K, Singh UB, Sahu A, Shukla R (2015) Bioremediation of heavy metals from soil and aquatic environment: an overview of principles and criteria of fundamental processes. Sustainability 7:2189–2212. doi:10.3390/su7022189

Duressa D, Soliman K, Taylor R, Senwo Z (2011) Proteomic analysis of soybean roots under aluminum stress. Int J Plant Genomics 2011:282531

Engelsberger WR, Schulze WX (2012) Nitrate and ammonium lead to distinct global dynamic phosphorylation patterns when resupplied to nitrogen-starved Arabidopsis seedlings. Plant J 69:978–995

Ezaki B, Gardner RC, Ezaki Y, Matsumoto H (2000) Expression of aluminum-induced genes in transgenic Arabidopsis plants can ameliorate aluminum stress and/or oxidative stress. Plant Physiol 127:918–927

Flora SJ, Pachauri V (2010) Chelation in metal intoxication. Int J Environ Res Public Health 7:2745–88

Fukao Y, Ferjani A, Tomioka R, Nagasaki N, Kurata R, Nishimori Y (2011) iTRAQ analysis reveals mechanisms of growth defects due to excess zinc in Arabidopsis. Plant Physiol 155:1893–1907

Gao XP, Flaten DN, Tenuta M, Grimmett MG, Gawalko EJ, Grant CA (2011) Soil solution dynamics and plant uptake of cadmium and zinc by durum wheat following phosphate fertilization. Plant Soil 338:423–434

Gill SS, Gill R, Anjum NA (2014) Target osmoprotectants for abiotic stress tolerance in crop plants—glycine betaine and proline. In: Anjum NA, Gill SS, Gill R (eds) Plant adaptation to environmental change: significance of amino acids and their derivatives. CAB International, Wallingford, pp 97–108

Goupil P, Souguir D, Ferjani E, Faure O, Hitmi A, Ledoigt G (2009) Expression of stress-related genes in tomato plants exposed to arsenic and chromium in nutrient solution. Plant Phys 166:1446–1452

Guo TR, Zhang GP, Zhou MX, Wu FB, Chen JX (2004) Effect of aluminum and cadmium toxicity on growth and antioxidant enzyme activities of two barley genotypes with difference Al tolerance. Plant Soil 258(1):241–248

Guo J, Dai X, Xu W, Ma M (2008) Overexpressing *GSH1* and *AsPCS1* simultaneously increases the tolerance and accumulation of cadmium and arsenic in *Arabidopsis thaliana*. Chemosphere 72:1020–1026

Gustin JL, Loureiro ME, Kim D, Na G, Tikhonova M, Salt DE (2009) MTP1-dependent Zn sequestration into shoot vacuoles suggests dual roles in Zn tolerance and accumulation in Zn-hyperaccumulating plants. Plant J 57:1116–1127

Hall J (2002) Cellular mechanisms for heavy metal detoxification and tolerance. J Exp Bot 53:1–11

Hao HZ, Chen TB, Jin MG, Lei M, Liu CW, Zu WP, Huang LM (2011) Recent advance in solidification/stabilization technology for the remediation of heavy metals-contaminated soil. Ying Yong Sheng Tai Xue Bao 22:816–824

Hasan SA, Fariduddin Q, Ali B, Hayat S, Ahmad A (2009) Cadmium: toxicity and tolerance in plants. J Environ Biol 30:165–174

Hasan SA, Hayat S, Ali B, Ahmad A (2008) 28-homobrassinolide protects chickpea (Cicer arietinum) from cadmium toxicity by stimulating antioxidants. Environ Pollut 151(1):60–66

Hassinen V, Vallinkoski VM, Issakainen S, Tervahauta A, Karenlampi S, Servomaa K (2009) Correlation of foliar MT2b expression with Cd and Zn concentrations in hybrid aspen (*Populus tremula × tremuloides*) grown in contaminated soil. Environ Pollut 157:922–930

Hassinen VH, Tervahauta AI, Schat H, Karenlampi SO (2011) Plant metallothioneins-metal chelators with ROS scavenging activity? Plant Biol 13:225–232

Hazrat A, Ezzat K, Muhammad AS (2013) Phytoremediation of heavy metals—concepts and applications. Chemosphere 91(7):869–881

Hong SP, Kim MJ, Jung MY, Jeon H, Goo J, Ahn SK (2008) Biopositive effects of low dose UVB on epidermis: coordinate upregulation of antimicrobial peptides and permeability barrier reinforcement. J Invest Dermatol 128:2880–2887

Hossain MA, Piyatida P, da Silva JAT, Fujita M (2012) Molecular mechanism of heavy metal toxicity and tolerance in plants: central role of glutathione in detoxification of reactive oxygen species and methylglyoxal and in heavy metal chelation. J Bot 2012:872–875

Hussain S, Siddique T, Arshad M, Saleem M (2009) Bioremediation and phytoremediation of pesticides: recent advances. Crit Rev Environ Sci Technol 39:843–907

Iglesia-Turino S, Febrero A, Jauregui O, Caldelas C, Araus JL, Bort J (2006) Detection and quantification of unbound phytochelatin 2 in plant extracts of Brassica napus grown with different levels of mercury. Plant Physiol 142(2):742–749

Jia Y, Tang SR, Wang RG, Ju XH, Ding YZ, Tu SX, Smith DL (2010) Effects of elevated CO_2 on growth, photosynthesis, elemental composition, antioxidant level, and phytochelatin concentration in *Lolium mutiforum* and *Lolium perenne* under Cd stress. J Hazard Mater 180(1–3):384–394

Kawachi M, Kobae Y, Mimura T, Maeshima M (2008) Deletion of a histidine-rich loop of AtMTP1, a vacuolar Zn^{2+}/H^+ antiporter of Arabidopsis thaliana, stimulates the transport activity. J Biol Chem 283:8374–8383

Klaassen CD (2006) Heavy metals and heavy metal antagonists. In: Goodman L, Gilman A (eds) The pharmacological basis of therapeutics. Medical Publishing Division, New York, pp 1825–1872

Kobae Y, Sekino T, Yoshioka H, Nakagawa T, Martinoia E, Maeshima M (2006) Loss of AtPDR8, a plasma membrane ABC transporter of Arabidopsis thaliana, causes hypersensitive cell death upon pathogen infection. Plant Cell Physiol 47:309–318

Konotop Y, Mezsaros P, Matusikova I, Batsmanova L, Taran N (2012) Application of nitrogen nutrition for improving tolerance of soybean seedlings to cadmium. Environ Exp Biol 10:139–144

Korenkov V, Park SH, Cheng NH, Sreevidya C, Lachmansingh J, Morris J, Hirschi K, Wagner GJ (2007) Enhanced Cd^{2+} selective root-tonoplast-transport in tobaccos expressing Arabidopsis cation exchangers. Planta 225:403–411

Kovacs E, Nyitrai P, Czovek P, Ovari M, Keresztes A (2009) Investigation into the mechanism of stimulation by low-concentration stressors in barley seedlings. J Plant Physiol 166:72–79

Kramer U, Pickering IJ, Prince RC, Raskin IL, Salt DE (2000) Subcellular localization and speciation of nickel in hyperaccumulator and non-accumulator Thlaspi species. Plant Physiol 122:1343–1353

Lanquar V, Ramos MS, Lelièvre F, Barbier-Brygoo H, Krieger-Liszkay A, Krämer U, Thomine S (2010) Export of vacuolar manganese by AtNRAMP3 and AtNRAMP4 is required for optimal photosynthesis and growth under manganese deficiency. Plant Physiol 152:1986–1999

Leitenmaier B, Kupper H (2011) Cadmium uptake and sequestration kinetics in individual leaf cell protoplasts of the Cd/Zn hyperaccumulator Thlaspi caerulescens. Plant Cell Environ 34:208–219

Li Y, Wei, Chen J (2004) Glutathione: a review on biotechnological production. Appl Microbiol Biotechnol 66:233–242

Lin W, Xiao T, Wu Y, Ao Z, Ning Z (2012) Hyperaccumulation of zinc by Corydalis davidii in Zn-polluted soils. Chemosphere 86(8):837–842

Liu Y, Jiang H, Zhao Z, An L (2011) Abscisic acid is involved in brassinosteroids-induced chilling tolerance in the suspension cultured cells from chorispora bungeana. J Plant Physiol 168(9):853–862

Llamas A, Ullrich CU, Sanz A (2008) Ni^{2+} toxicity in rice: effect on membrane functionality and plant water content. Plant Physiol Biochem 46:905–10

Ma X, Richter AR, Albers S, Burken JG (2005) Phytoremediation of MTBE with hybrid poplar trees. Int Phytoremediation 6(2):157–167

Madejczyk MS, Ballatori N (2012) The iron transporter ferroportin can also function as a manganese exporter. Biochim Biophys Acta Biomembr 1818(3):651–657

Maksymiec W, Wianowska D, Dawidowicz AL, Radkiewicz S, Mardarowicz M, Krupa Z (2005) The level of jasmonic acid in Arabidopsis thaliana and Phaseolus coccineus plants under heavy metal stress. J Plant Physiol 162:1338–1346

Marentes E, Rauser WE (2007) Different proportions of cadmium occur as Cd-binding phytochelatin complexes in plants. Physiol Plant 131:291–301

Masarovicova E, Kralova K, Kummerova M (2010) Principles of classification of medicinal plants as hyperaccumulators or excluders. Acta Physiol Plant 32:823–829

Masarovicova E, Kralova K, Sersen F (2011) Plant responses to toxic metal stress. In: Pessarakli M (ed) Handbook of plant and crop stress, 3rd edn. CRC Press, Boca Raton, pp 595–634. ISBN 978-1-4398-1396-6

Masood A, Iqbal N, Khan NA (2012) Role of ethylene in alleviation of cadmium-induced photosynthetic capacity inhibition by sulphur in mustard. Plant Cell Environ 35:524–533

Mench M, Martin E (1991) Mobilization of cadmium and other heavy metals from two soils by root exudates of Zea mays L., Nicotiana tabacum L and Nicotiana rustica L. Plant and Soil 132:187–196

Menezes-Benavente L, Teixeira FK, Kamei CLA, Margis-Pinheiro M (2004) Salt stress induces altered expression of genes encoding antioxidant enzymes in seedlings of a Brazilian indica rice (*Oryza sativa*L.). Plant Sci 166:323–331

Metwally A, Finkemeier I, Georgi M, Dietz K-J (2003) Salicylic acid alleviates the cadmium toxicity in barley seedlings. Plant Physiol 132:272–281

Miguel AMR, Ondrej Z, Marketa K, Vojtech A, Miroslava B, Rene K (2013) Analysis of cadmium-phytochelatins 2 complexes using flow injection analysis coupled with electrochemical detection mass spectrometry. Int J Electrochem Sci 8:4409–4421

Miller G, Shulaev V, Mittler R (2008) Reactive oxygen signalling and abiotic stress. Physiol Plant 133:481–489

Mohanpuria P, Rana NK, Yadav SK (2007) Cadmium induced oxidative stress influence on glutathione metabolic genes of *Camellia sinensis* (L.) O. Kuntze. Environ Toxicol 22:368–374

Mostofa MG, Hossain MA, Fujita M, Tran LSP (2015) Physiological and biochemical mechanisms associated with trehalose-induced copperstress tolerance in rice. Sci Rep 5:11433

Moulick A, Mukhopadhyay D, Talapatra S, Ghoshal N, Raychaudhuri SS (2013) Molecular cloning, modeling, and characterization of type 2 metallothionein from Plantago ovata Forsk. Sequencing 2013:10

Nagajyoti P, Lee K, Sreekanth T (2010) Heavy metals, occurrence and toxicity for plants: a review. Environ Chem Lett 8:199–216

Nagasaka S, Takahashi M, Itai RN, Bashir K, Nakanishi H, Mori S, Nishizawa NK (2009) Time course analysis of gene expression over 24 hours in Fe–deficient barley roots. Plant Mol Biol 69(5):621–631

Naser HM, Sultana S, Mahmud NU, Gomes R, Noor S (2011) Heavy metal levels in vegetables with growth stage and plant species variations. Bangladesh J Agril Res 36(4):563–574

Navarro CA, Orellana LH, Mauriaca C, Jerez CA (2009) Transcriptional and functional studies of*Acidithiobacillus ferrooxidans* genes related to survival in the presence of copper. Appl Environ Microbiol 75:6102–6109

Ngu TT, Dryden MD, Stillman MJ (2010) Arsenic transfer between metallothionein proteins at physiological pH. Biochem Biophys Res Commun 401:69–74

Nishiyama R, Watanabe Y, Leyva-Gonzalez MA, Ha CV, Fujita Y, Tanaka M (2013) *Arabidopsis* AHP2, AHP3, and AHP5 histidine phosphotransfer proteins function as redundant negative regulators of drought stress response. Proc Natl Acad Sci U S A 110:4840–4845

Noctor G, Mhamdi A, Chaouch S, Han Y, Neukermans J, Marquez-Garcia B et al (2012) Glutathione in plants: an integrated overview. Plant Cell Environ 35:454–484

Ortega-Villasante C, Hernández LE, Rellan-Alvarez R, del Campo FF, Carpena-Ruiz RO (2007) Rapid alteration of cellular redox homeostasis upon exposure to cadmium and mercury in alfalfa seedlings. New Phytol 176:96–107

Ortega-Villasante C, Rellan-Álvarez R, Del Campo FF, CarpenaRuiz RO, Hernandez LE (2005) Cellular damage induced by cadmium and mercury in Medicago sativa. J Exp Bot 56:2239–2251

Park J, Song WY, Ko D, Eom Y, Hansen TH, Schiller M, Lee TG, Martinoia E, Lee Y (2012) The phytochelatin transporters AtABCC1 and AtABCC2 mediate tolerance to cadmium and mercury. Plant J 69:278–288

Pesko M (2009) Energetic plants—cost and benefit & Masarovicova E, Kralova K. Ecol Chem Eng 16(3):263–276, ISSN 1231–7098

Rausch T, Gromes R, Liedschulte V, Muller I, Bogs J, Galovic V, Wachter A (2007) Novel insight into the regulation of GSH biosynthesis in higher plants. Plant Biol 9:565–572

Rea PA (2007) Plant ATP-binding cassette transporters. Annu Rev Plant Biol 58:347–375

Rellan-Alvarez R, Abadia J, Alvarez-Fernandez A (2008) Formation of metal-nicotianamine complexes as affected by pH, ligand exchange with citrate and metal exchange. A study by electrospray ionization time-of-flight mass spectrometry. Rapid Commun Mass Spectrom 22:1553–1562

Remans T, Opdenakker K, Smeets K, Mathijsen D, Vangronsveld J, Cuypers A (2010) Metal-specific and NADPH oxidase dependent changes in lipoxygenase and NADPH oxidase gene expression in Arabidopsis thaliana exposed to cadmium or excess copper. Funct Plant Biol 37:532–544

Rojas LA, Yanez C, Gonzalez M, Lobos S, Smalla K, Seeger M (2011) Characterization of the metabolically modified heavy metal-resistant Cupriavidus metallidurans strain MSR33 generated for mercury bioremediation. PLoS One 6:e17555

Samaranayake P, Peiris BD, Dssanayake S (2012) Effect of excessive ferrous (Fe^{2+}) on growth and iron content in rice (*Oryza sativa*). Int J Agri Biol 14:296–298

Sarwar N, Saifullah MSS, Zia MH, Naeem A, Bibi S, Farida G (2010) Role of mineral nutrition in minimizing cadmium accumulation by plants. J Sci Food Agric 90:925–937

Sato Y, Masuta Y, Saito K, Murayama S, Ozawa K (2011) Enhanced chilling tolerance at the booting stage in rice by transgenic overexpression of the ascorbate peroxidase gene, OsAPXa. Plant Cell Rep 30:299–406

Satofuka H, Fukui T, Takagi M, Atomi H, Imanaka T (2001) Metal-binding properties of phytochelatin-related peptides. J Inorg Biochem 86:595–602

Savci S (2012) An agricultural pollutant: chemical fertilizer. Int J Environ Sci Dev 3:77–80

Schellingen K, Van Der Straeten D, Vandenbussche F, Prinsen E, Remans T, Vangronsveld J, Cuypers A (2014) Cadmium-induced ethylene production and responses in Arabidopsis thaliana rely on ACS2 and ACS6 gene expression. BMC Plant Biol 14:214

Shamsi IH, Wei K, Zhang GP, Jilan GH, Hassan MJ (2008) Interactive effects of cadmium and aluminum on growth and antioxidative enzymes in soybean. Biolog Plant 52(1):165–169

Takahashi F, Yoshida R, Ichimura K, Mizoguchi T, Seo S, Yonezawa M, Maruyama K, Yamaguchi-Shinozaki K, Shinozaki K (2007) The mitogen-activated protein kinase cascade MKK3-MPK6 is an important part of the jasmonate signal transduction pathway in Arabidopsis. Plant Cell 19:805–818

Troadec MB, Ward DM, Lo E, Kaplan J, De Domenico I (2010) Induction of FPN1 transcription by MTF-1 reveals a role for ferroportin in transition metal efflux. Blood 116:4657–4664

Vander EA, Baker AJM, Reeves RD, Pollard AJ, Schat H (2013) Hyperaccumulators of metal and metalloid trace elements: facts and fiction. Plant Soil 362:319–34

Verret F, Gravot A (2004) Overexpression of AtHMA4 enhances root-to-shoot translocation of zinc and cadmium and plant metal tolerance. FEBS Lett 576(3):306–312

Vert G, Grotz N, Dedaldechamp F, Gaymard F, Guerinot ML, Briat JF (2002) IRT1, and Arabidopsis transporter essential for iron uptake from the soil and for plant growth. Plant Cell 14:1223–12233

Vinita H (2007) Phytoremediation of toxic metals from soil and waste water. J Environ Biol 28(2):367–376

Wang J, Feng X, Anderson CWN, Xing Y, Shang L (2012) Remediation of mercury contaminated sites—a review. J Hazard Mater 221(222):1–18

Wang Y, Stass A, Horst WJ (2004) Apoplastic binding of aluminum is involved in silicon-induced amelioration of aluminum toxicity in maize. Plant Physiol 136:3762–70

Weber M, Harada E, Vess C, von Roepenack-Lahaye E, Clemens S (2004) Comparative microarray analysis of Arabidopsis thaliana and Arabidopsis halleri roots identifies nicotianamine synthase, a ZIP transporter and other genes as potential metal hyperaccumulation factors. Plant J 37:269–281

Webber MB, Schat H, Ten Bookum-Van Der Maarel WM (1991) The effect of copper toxicity on the contents of nitrogen compounds in Silene vulgaris (monech.) Garacke. Plant soil 133:101–109

Xiang C, Werner BL, Christensen EM, Oliver DJ (2001) The biological functions of glutathione revisited in Arabidopsis transgenic plants with altered glutathione levels. Plant Physiol 126:564–574

Xie QE, Yan XL, Liao XY, Li X (2009) The arsenic hyperaccumulator fern Pteris vittata L. Environ Sci Technol 43:8488–8495

Xue TT, Li XZ, Zhu W, Wu CG, Yang GG, Zheng CC (2009) Cotton metallothionein GhMT3a, a reactive oxygen species scavenger, increased tolerance against abiotic stress in transgenic tobacco and yeast. J Exp Bot 60:339–349

Yadav SK (2010) Heavy metals toxicity in plants: an overview on the role of glutathione and phytochelatins in heavy metal stress tolerance of plants. S Afr J Bot 76:167–179

Yang G, Wang Y, Xia D, Gao C, Wang C, Yang C (2014) Overexpression of a GST gene (*ThGSTZ1*) from *Tamarix hispida* improves drought and salinity tolerance by enhancing the ability to scavenge reactive oxygen species. Plant Cell Tiss Organ Cult 117:99–112

Yin Z, Jiang H, Lee EY, Ni M, Erickson KM, Milatovic D, Bowman AB, Aschner M (2010) Ferroportin is a manganese-responsive protein that decreases manganese cytotoxicity and accumulation. Neurochemistry 112:1190–1198

Ying RR, Qiu RL, Tang YT, Hu PJ, Qiu H, Chen H-R (2010) Cadmium tolerance of carbon assimilation enzymes and chloroplast in Zn/Cd hyperaccumulator Picris divaricata. J Plant Physiol 167:81–7

Yu T, Li YS, Chen XF, Hu J, Chang X, Zhu YG (2003) Transgenic tobacco plants overexpressing cotton glutathione S-transferase (GST) show enhanced resistance to methyl viologen. J Plant Physiol 160(11):1305–11

Zhao C, Ikka T, Sawaki Y, Kobayashi Y, Suzuki Y, Hibino T (2009) Comparative transcriptomic characterization of aluminum, sodium chloride, cadmium and copper rhizotoxicities in Arabidopsis thaliana. BMC Plant Biol 9:32–47

Zhao C, Qiao M, Yu Y, Xia G, Xiang F (2010) The effect of the heterologous expression of *Phragmites australis* γ-glutamylcysteine synthetase on the Cd^{2+}accumulation of *Agrostis palustris*. Plant Cell Environ 33:877–887

Zuo Y, Zhang F (2011) Soil and crop management strategies to prevent iron deficiency in crops. Plant Soil 339:83–95

Heavy Metal Stress Signalling in Plants

3

Contents

Abstract

Plants tolerate diverse abiotic stresses during entire their life cycle. One among such abiotic stresses was heavy metal stress. Heavy metals generate diverse signals as they intermingle with other metabolic pathways. In this association, chemical reactivity displays fundamental role particularly among essential and nonessential metals. Though, such interface does not certainly have a dismal effect for the plant overall. As heavy metal recognizes heavy metal through receptor, signal transduction pathway commences. However, scarce literature presents evidences regarding primary recognition via receptor. The most apparent entrants responsible for heavy metal acquisition are plasma membrane proteins such as reductases and transporters. In addition to them, there are numerous other sensors arising from physical modifications in cellular structures stimulated by metal pollution. In case, metal is identified by cells that promote cellular signal transduction and they will utilize aspects of prevalent signalling pathways such as calcium fluxes. To facilitate stress signalling response, plant cells ought to identify these signals and switch them into an apt reaction, which consecutively impart on plants the capacity to tolerate adverse surroundings. The tolerance strategy necessitates synchronization of intricate physiological and biochemical pathways, together with variation in

S. Jan, J.A. Parray, *Approaches to Heavy Metal Tolerance in Plants*,
DOI 10.1007/978-981-10-1693-6_3

comprehensive gene expression, protein amendment, and primary and secondary metabolite symphony. In the preceding years, practical genomics advance has rather dilapidated the intricate mechanisms that constrain from stress perception and transduction, via trafficking of signal molecules, to the articulated amendment of genes accountable for plant stress response. Besides, the explication of the role of recently recognized stress-responsive noncoding RNA will assist perception of the intricate reaction to stress.

Keywords
Heavy metal stress • Signal transduction • Transcription factors • Genetic manipulation • Kinases

3.1 Introduction

Plants exhibit immense capacity to acclimatize their metabolism to instantaneous changes in the environment. To meet this rationale, they are armored with intricate processes, such as perception transduction and stress stimuli transmission (Solanki and Dhankar 2011; Khan et al. 2014; Islam et al. 2015). Plants acquire precise mechanism escalating their response to the stress stimulus. It provokes an apt riposte even in response to lesser intense stimuli, for instance, via deflation or pathogenesis, which advances intensity further to jeopardize their existence (Cabello et al. 2014). These kinds of mechanisms associated to external stimuli, their development in cells, involving organs and organisms consisting of biochemical processes enhance the action of specific signalling pathways (Rejeb et al. 2014). Overall, plant species possess varied types of enzymes implicated in biosynthesis of particular signalling molecules augmenting the stress reaction response (Wasternack and Hause 2013; Okazaki and Saito 2014). An indispensable fundamental of the signalling pathways illustrates an immense similitude across the world of flora and fauna, which is established by the incidence of receptors and/or ion channels, signalling molecules, and transcription factors specifically influencing the expression of genes via forming the foundation of adaptation or defense reaction of the organism (Oldroyd 2013). Diverse receptors have been identified, for instance, ethylene (Binder et al. 2012), systemin (Antunes et al. 2011), osmosensor-transmembrane histidine kinase (Osakabe et al. 2013), and putative receptor of abscisic acid (Erb et al. 2012). Like animals, plant also displays similar array of calmodulin-Ca^{2+} system which consequently activates kinases resulting in assorted biochemical pathways in plants (Means 2013; Walters 2015). This calmodulin–Ca^{2+} system establishes concert forming the electric potentials along with other signalling molecules which convey signals about stress state inside the cell and its adjoining cells. These signalling molecules are so closely associated that resulting pathways from several stress stimuli are arranged into web, for example, in response to abiotic stress salt, cold, and drought stress (Huang et al. 2012), mechanical wounding, and pathogenic attack (Arimura et al. 2011; Meng and Zhang 2013; Moghaddam et al. 2015). Heavy metal contamination is major abiotic stress currently prevailing over living organisms due to overurbanization and intensive industrialization. Heavy metals significantly inhibit growth and promote senescence. The mechanism underlying heavy metal-induced toxicity and its accumulation in diverse subcellular organelles and various approaches trailed by plants to counteract HM stress have been demonstrated comprehensively in preceding assessment, but key aspects at diverse signalling phases that direct to HM tolerance have not been premeditated appropriately.

Generally, HM contamination in plants is illustrated by production of stress proteins, signalling molecules, and chelators such as organic acids and GSH-mediated phytochelatins. However, tolerance strategy is metal specific, but toxicity mechanisms are communal. Similarly, cells may accumulate normal and explicit metal response to neutralize heavy metal toxicity. In this chapter we will illustrate advances made on molecular elements of heavy metal response and related regulatory pathways in plants, for example, genomic expression of sensors, cascades of kinases, cis-regulatory elements, transcription factors, and aquaporins/transporters to comprehend HM-stimulated signalling network and cross talk with allied stresses.

3.1.1 Direct Action of Heavy Metals

The phenomena of direct action of heavy metals and their impacts on plants have been illustrated previously (Hernandez et al. 2012). Perusal of literature and preceding in vivo and particularly in vitro investigations revealed heavy metals can furnish protein related with water splitting organization (Martin 2012), lipids particularly galactolipids associated with PSII, and elemental constituents of thylakoid membranes requisite for photosynthetic activity (Thapa et al. 2012; Sytar et al. 2013). The surplus of heavy metals can efficiently attach to membranes, particularly subsequent to exposition via oxygen atoms (Parmar et al. 2013) or histidine, tryptophan, and tyrosine group of polypeptides (Vandenbossche et al. 2014) and shows interference with electron flow between various electron donor and acceptor centers such as quinone of PS II, from Tyr Z to P_{680+}, and cyt b559 of PSII reaction center (Vass 2012). Heavy metal has deleterious effects pertaining to chlorophyll structure like substitution of Mg by heavy metal in Chl (Nakajima et al. 2013). Moreover, Chl synthesis can further be hampered by malfunction of enzymes requisite for Chl synthesis (Parmar et al. 2013). Research has confirmed the substitution of Hg by Cu in plastocyanin, thereby obstructing the electron flow to PSI (Hernandez et al. 2012). HM stress modulates gene expression manifested through diverse signalling surges (DalCorso et al. 2010). Proteomic research has further advanced our understanding pertaining to heavy metal responses. One of the major responses to heavy metal stress is displayed through the disequilibrium in Ca ions at cellular levels and associated protein calmodulin which maintains ionic homeostasis and enhances gene expression adding to metal tolerance signifying association with cross talks (Gallego et al. 2012; Poovaiah et al. 2013). Moreover, the decreased levels of glutathione raise following heavy metal exposure with equivalent upregulation of antioxidant enzymes and genes responsible for its upregulation (Gill et al. 2013). Cross-talk flanking among biotic and heavy metal stress is manifested through accretion of jasmonic acid, ethylene, and salicylic acid (SA) HM stressed roots (Popova et al. 2012; Thao et al. 2015). Heavy metal stress and ROS-induced phytoalexin biosynthesis when exposed to heavy metal stress signify common signalling pathways (Smékalová et al. 2014). However, HM signalling is metal specific. Heavy metal signalling allocates signalling pathway in exposure to NO elevation for plant defense (Cerana and Malerba 2015). Heavy metal tolerance entails increased Ca levels, calmodulin and MAPK activation, GSH-mediated regulation, and ROS and NO signalling that confer improved tolerance strategy.

3.1.2 Indirect Action of Heavy Metals

Heavy metal stress induced mineral nutrient disturbances consequently affecting plant growth (Singh et al. 2011). Alaoui-sosse et al. (2004) illustrated copper-induced nutrient disturbances and decrease in photosynthetic rate via sugar accumulation. Excessive Cd accumulation affects the uptake and distribution of some nutrients in the potato cultivars and therefore may be dependable for mineral disturbances and depression of plantlet growth (Nazar et al. 2012). Decrease in the mitotic index was observed in the case of Pb, Hg, as well as Cd ions (Barbosa et al. 2010).

Parween et al. (2011) demonstrated *Vicia faba* L. seeds grown in the nutrient medium for 48 h containing different concentrations of cadmium chloride (control, 0.05 mM, 0.1 mM, 0.2 mM, 0.4 mM, and 0.8 mM) showed different genotoxic effects such as polyploidy, multipolarity, chromosomal bridge with fragments, lagging chromosome, and micronuclei. Guzel and Terzi (2013) reported excess copper caused diminution in the growth; leaf water potential; Na^+, K^+, Ca^+, and Mg^{2+} concentrations; and soluble protein levels but increment in proline, total soluble sugars, and Cu^{2+} contents. Sublethal and half of sublethal concentration of cadmium and copper in *Cleome gynandra* showed increased accumulation of total phenolics, percent antioxidant activity, and enzymatic antioxidants, thereby increasing the tolerance of plant to cope with oxidative stress (Haribabu and Sudha 2011). Anjum et al. (2012) reported that toxic metals and metalloids alter differentially the total GSH, its redox couple (GSH/GSSG), and associated main enzymes (such as, GR, GPXs, GSTs) in plants which consecutively, cumulatively control plants' adaptation and tolerance to TMs and metalloids stress. Heavy metal accumulation induces growth inhibition, senescence stimulation, and shortening and thickening of roots along with many other physiological and biochemical disorders. Manikandan et al. (2015) reported that the application of low concentration of $CuCl_2$ salt exhibited significant increment in defense-related enzyme accumulation like PO, PPO, PAL, β-1,3 glucanases, and total phenol in *Mentha arvensis* L. seedlings. Cadmium accumulation enhances NADPH oxidase and NADPH dehydrogenase activity in cucumber seedlings (Jakubowska et al. 2015). Mota et al. (2015) illustrated effects of numerous heavy metals on growth, exopolysaccharide production, ultrastructure, and protein profiles of cyanobacterium *Cyanothece* sp. CCY 0110. Hediji et al. (2015) reported significant decrease in K and Mg content in all the plants organs of *Solanum lycopersicum* L., thereby inducing severe nutritional disturbances leading to fruit abortion. Srivastava et al. (2015) demonstrated toxic effects of mercury in plants manifested through abscission of older leaves, declined growth, and reduced vigor.

3.2 Hormone Signalling Pathways

During the past 20 years, it has been evaluated that heavy metal stress stimulates phytochelatins, metallothioneins, and specific stress protein synthesis (Sunitha et al. 2013). It is highly intricate issue to verify if the deleterious effects following heavy metal exposure were the outcome of signalling pathways developing via interplay of produced chemicals or analogous through heavy metal action and signalling pathway. The preliminary evidence that specified the effect of signalling modules in response to copper stress which induced ethylene synthesis was confirmed by Sandmann and Boger (1980). This study demonstrated enhanced ethylene production in response to copper toxicity leading to decline in photosystem activity consequently leading to senescence processes (Jonak et al. 2004). Increased ethylene levels under stress environment can also enhance rigidity of cell walls via intensification of its lignification mechanism subsequent to growth inhibition (Potters et al. 2007). Copper toxicity induced increased ethylene which consequently leads to decline in plant growth (Li and Li 2004). Ethylene production increases in response to Cd, Cu, Fe, and Zn, but Co has no effect over ethylene levels (Arteca and Arteca 2007). Ethylene production increased as result of Cu and cadmium exposure through enhanced gene expression of enzyme which produces ACC synthase (Abeles et al. 2012). Steffens (2014) reported increased ethylene production following Cu and Zn stress along with high lipoxygenase activity. It signifies that heavy metals can stimulate the jasmonate pathway and/or lipoxygenase-mediated reactive oxygen species (ROS) formation leading to increased ethylene concentration (Thao et al. 2015), by enhancing ACC synthase and oxidase activity (Glick 2015).

Hanaka et al. (2016) demonstrated implication of methyl jasmonate (MJ) as a signal molecule which functions in tolerance strategy against copper (Cu)-induced oxidative in the *Phaseolus coccineus* plants. Continuous exposure to Cu or Cd leads to biphasic jasmonic acid accretion resulting in rapid decline of growth and abrupt senescence of plants (Fiala et al. 2015). This observation further confirmed though jasmonates' function as signal molecules their conferring tolerance to stress; they are also responsible for decreased photosynthetic activity and decline in growth (Sirhindi et al. 2015) and rapid advancement toward senescence (Dar et al. 2015). Xiao et al. (2004) confirmed implication of COI1 gene-encoding proteins belonging to SCFCOI1-type complex in JA-induced decline in root growth and riboflavin content. Zhang et al. (2001) reported enhanced expression of specific gene like *PvSR2* prerequisite plant tolerance. Induced expression of genes responsible for heavy metal tolerance via signalling modules such as jasmonates, salicylic acid, ethylene, and abscisic acid is stimulated through biotic and abiotic tress factors (Loake and Grant 2007). Jozefczak et al. (2012) demonstrated general reaction of Cd/Cu ions and JA in gene transcription implicated in glutathione metabolism. Similarly induction of transcription in specific genes like VSP2 (vegetative-storage protein) and MAPK (mitogen-activated protein kinase) transcript further signifies the heavy metal effect via signalling pathway (Hernandez et al. 2012). In lieu with this, it can be speculated that heavy metal contamination at lower concentration could provoke signalling pathways which confer tolerance to plants. Numerous such studies have been reported where low heavy metal stress like cadmium can confer high tolerance against biotic stress (Mittra et al. 2004). Poschenrieder et al. (2006) further verified cross talk between biotic stress and heavy metal stress and stimulated varied defense pathways via activation of genes that regulate enzyme activity of glucanases, chitinases, or proteinases (Ovecka and Takac 2014).

3.2.1 Signalling Through Reactive Oxygen Species (ROS)

Oxidative stress is signified by enhancement in genetic expression of superoxide anion radical (O_2^-) content, H_2O_2, and hydrogen peroxide radical (OH^-) persistently formed throughout plant growth (Schmitt et al. 2014). An increase in enzyme activity and inducing pathways that generate H_2O_2 and O_2^- and activate the function of antioxidative systems leads to generation of OH^- radical (Circu and Aw 2010). Heavy metals like Fe^{2+} and Cu^+ generally react with H_2O_2 and OH^- through Haber–Weiss reaction. The intense effect of heavy metals on oxidative defense system is fundamental for their association with allied signalling response. Karuppanapandian et al. (2011) demonstrated increased ROS generation via heavy metals' action annihilates cellular constituents, for instance, membranes, nucleic acids, and chloroplast pigments. The generation of lipid peroxides is a result of heavy metal-induced oxidative stress and may possibly function as an activation signal for plant defense genes via high throughput of octadecanoid pathways (Zhang et al. 2008). Increased production of thiobarbituric acid (TBA) content in leaves in response to Cu stress has been demonstrated (Posmyk et al. 2009). Popova et al. (2009) demonstrated amelioration of Cd toxicity via salicylic acid (SA), as an identified signalling factor that frequently hampers the jasmonate (JA) pathways. Sirhindi et al. (2015) have demonstrated that copper induces H_2O_2-dependent peroxidase activity pursued by cell wall rigidity owing to cross-linkage between the cell wall polymers. Cadmium leads to accretion of H_2O_2 and decline in growth exhibited following 1 day of metal contamination (Weckx and Clijsters 1996) or immediately following the first hour (Maksymiec and Krupa 2006); inference of signal molecules, together with H_2O_2 categorized as signalling factor, is rather feasible. Takac et al. (2011) demonstrated that higher H_2O_2 formation, particularly via enhanced NADPH oxidase activity, can decline cell wall extensibility. NADPH oxidase is implicated in plant metabolism and their reaction to numerous biotic stresses in addition to Cu (Remans et al. 2010). Suspension of culture cells under cadmium contamination exhibited decline in growth followed by lower expression of cyclin B1 (Sobkowiak and Deckert 2004). The status of cellular redox system is usually manifested through higher accretion of H_2O_2 (Jaspers and Kangasjärvi 2010) and may improve the plant redox status via calcium signalling in the genetic expression of glutathione transferase gene (Szalai et al. 2009). Conversely, H_2O_2 like jasmonate resulted in closure of stomata, and Calvin cycle can consequently decline as was seen following Cu or Cd stress (Gill and Tuteja 2010). Higher H_2O_2 accretion has been exhibited from jasmonate activity signifying intense association among the production of the signal molecules (Sharma et al. 2012). Conversely, the accessible resources are ambiguous if heavy metals function on oxidative processes or/and via initiation of the signalling pathways (Cuypers et al. 2012). H_2O_2 accretion has been exhibited in response to Cu, Cd, and Hg contamination in *A. thaliana* (Opdenakker et al. 2012) and tomato plants (Cuypers et al. 2011) correspondingly. Conversely, in barley plants only Mn enhanced the H_2O_2 levels following 5 days of exposure (Demirevska-Kepova et al. 2004). This discrepancy might signify that H_2O_2 accretion extended diversely throughout prolonged stress. The system of heavy metal-induced H_2O_2 accretion is not currently elucidated. Even though some of research studies emphasize the precise function of CAT and SOD and/or higher O_2^- generation in this mechanism that the primary effect of numerous heavy metals might be the consequence of higher NADPH oxidase activity moderately associated with higher O_2^- generation (Sun et al. 2016) and jasmonate pathway action with higher SOD activity (Cao et al. 2009). Conversely, this impact was associated with reduction of the enzymic antioxidative system, and higher lipid peroxidation might result as H_2O_2 content decline. The period for H_2O_2 incidence in 1 h displayed the following Cu or Cd contamination was nearer to that of other allied stress factors in which signal modules were implicated. The brief exposure to H_2O_2 cannot destroy the cell but might be adequate to stimulate other considerable reactions, e.g., decline in cell expansion displayed following heavy metal exposure. It has been highly intriguing issue if jasmonate accumulated inside *A. thaliana* leaves immediately following 1 h heavy metal exposure, stimulating H_2O_2 generation, or if it accumulates following 6–7 h of growth with Cd or Cu exposure (Maksymiec and Krupa 2006). In ginseng cells exposed to oligogalacturonic acid as elicitor, H_2O_2 activates jasmonate accretion (Hu et al. 2003). Moreover, the result of H_2O_2 occurrence in the cells may be improvement of detrimental impact on chromatin of several heavy metals varying the oxidative level (Sharma and Dietz 2009). Cadmium increases the genotoxicity in plants as in mammalian cells by reticence of DNA repair and methylation mechanism (Bertin and Averbeck 2006). These research findings' results have indicated impacts of heavy metals on plants as a consequence of their unswerving impact on membranes and the photosynthetic apparatus and meandering effect rooted via stimulation of some signalling pathways (Gill and Tuteja 2011). The enhanced production of secondary metabolites and pathogenesis-associated proteins related with higher energy utilization might upshot in decline of primary metabolites and plant yield (Bartwal et al. 2013). Hence the tolerance strategy adopted by plant can be determined by the developmental stage, plant species, and time span for specific heavy metal exposure. The increased JA and H_2O_2 production might be due to heavy metal-stimulated disruption in the membranes and overall decline in water levels across the cells (Emamverdian et al. 2015). The effect of the antecedent of jasmonic acid–12-oxo-phytodienoic acid (OPDA), an effectual regulator of defense responses, in heavy metal stress is currently entirely unidentified and might in prospect elucidate the disparity in the mechanism among endogenously produced and exogenously sprayed jasmonate (Wasternack 2014). Further, the function of diverse vigorous plant oxylipins originated from linolenic acids and produced through jasmonate biosynthesis should be elucidated to precisely and comprehended how heavy metals function in whole concert. Precisely, the acquired pioneering records signify the prospect of circuitous heavy metal mechanism via stimulation of signalling pathways, particularly those associated with jasmonate, ethylene, and H_2O_2 (Hernández et al.2012). Signal transduction pathways involved via heavy metals might consequently pave toward rapid decline of growth mechanism (generally rising their tolerance to stress) or enhance senescence following a prolonged heavy metal/stress exposure.

3.3 Review of Abiotic Stress Features Generating MAPK Activity

Plants being sessile are frequently under diverse biotic and abiotic stresses in their surroundings. To combat this stress, plants acquire system for instant signal perception from the variable environmental conditions, and these transmit signals are transmitted via explicit adaptive/defensive pathways. Cristina et al. (2010) demonstrated the imperative role of mitogen-activated protein kinase (MAPK) pathways in signal transduction implicated in the maintenance of normal development and metabolism of cells under stress conditions. Lee et al. (2016) confirmed the activation of mitogen-activated protein kinase (MAPK) pathways following assorted array of abiotic stresses as in the case where *Arabidopsis* MPK4 and MPK6 are activated in response to salt, cold, drought, touch, and wounding (Nakagami et al. 2005). MAPK module implicated in signal transduction element has been identified in *Arabidopsis* under cold and salt stresses. This module consists of the MEKK1 as an upstream activator of MKK2 and the downstream MAPKs as MPK4 and MPK6 (Gao et al. 2008). Moreover, also MKK1 is activated via salt, drought, and wounding stress and phosphorylated to MPK4 which leads to its imperative role in signalling cascade against abiotic stress (Qiu et al. 2008). MPK3, MPK4, and MPK6 have exhibited enhanced expression in plantlets of *Arabidopsis* growing in cell suspensions under hyposmolarity (Droillard et al. 2004).

Increased levels of ozone triggered MAPK signalling pathways via intense ROS generation and accretion of ethylene, jasmonic acid (JA), and salicylic acid (SA), leading to local programed cell death (Pitzschke and Hirt 2009). MPK3 and MPK6 have been further triggered through ozone, leading to nuclear translocation of MAPKs in *Arabidopsis* (Liu et al. 2010). This activation is not dependent on ethylene and JA; however, MPK3 is triggered via salicylic acid (Kazan 2015). Recently, Smékalová et al. (2014) illustrated significant function of MKP2 (MAPK phosphatase 2) as an upregulator of the cellular response against ozone because it modulates the triggering phase of MPK3 and MPK6. Decline in MKP2 produces hypersensitivity in response to ozone with elaborate MPK3 and MPK6 activation process (Lei et al. 2014). NtMPK4 has also been reported to function in tobacco as a significant factor for ozone sensitivity and JA signalling (Kumar et al. 2013). Following this research investigation, transgenic plants illustrated NtMPK4 as significant kinases against wounding and have been associated with ozone tolerance via diverse stomatal closure patterns. Oxidative stress stimulated via application of H_2O_2 can also activate MPK1 and MPK2 (Zhou et al. 2014), MPK3 and MPK6 (Wang et al. 2010), MPK4 (Pitzschke et al. 2009), and MPK7 (Doczi et al. 2007) in *Arabidopsis*, indicating ROS control upstream of several MAPK cascades. MAPKs can also be activated via above toxic levels of heavy metals. Xie et al. (2014) demonstrated activation of MAPK cascade against cadmium and copper tress leading to implication of OsMAPK3 and OsMPK6 in signalling pathway of rice (Xie et al. 2014). Furthermore, triggering of four different MAPKs, for example, SIMK, MMK2, MMK3 and SAMK, was displayed following copper or cadmium stress in *Medicago sativa* seedlings (Jonak et al. 2004). Conversely, various MAPK pathways are likely to be involved in the defense against copper and cadmium stress.

Defense system against biotic stress works on two defense lines; the primary stage implicates precise recognition of conserved pathogen-associated molecular patterns (PAMPs) with transmembrane pattern recognition receptors

(PRRs) and triggers defense responses of plant cell (Boller and Felix 2009). These defense systems involve variation in enzymatic functions, gene expression reprogramming, and release of antimicrobial modules which consequently trigger PAMP-triggered immunity (PTI) that can hamper plant colonization via pathogen (Zipfel 2009). Pathogens counteract PTI by suppressing the PTI signalling modules through generation of effector proteins (Zhang and Zhou 2010). Following the secondary line of defense, plants differentiate pathogen-induced effector proteins via resistance (R) proteins. Resistance proteins act as immune receptors and activate host resistance by means of specified cell death confining the pathogen multiplicity. Such kind of plant immunity is termed as effector-triggered immunity (ETI) which is an enhanced and improved PTI response conferring disease resistance (Qi et al. 2011). During recent years it has been confirmed that MAPKs demonstrate imperative function in innate immune response and impart resistance against pathogens in *Arabidopsis*, rice, tobacco, parsley, tomato, and maize. In this context, the superlative characterized *Arabidopsis* MAPKs are MPK3, MPK4, and MPK6 which exhibited activation through bacterial and fungal PAMPs through plant-pathogen interactions (Lumbreras et al. 2010). Thus, such synchronized complex of MAPK cascades represented the fundamental phase of plant innate immunity involved in defense to microbial pathogens and their effectors along with surplus of abiotic stress reactions. *Arabidopsis* FLS2 (flagellin sensitive 2) is an extensively categorized plant PRR from the leucine-rich repeat receptor kinase family which is requisite for flagellin signalling (Boller and He 2009). Sörensson et al. (2012) illustrated elicitation through bacterial elicitor flagellin released by eubacteria leads to internalization of plasma membrane receptor FLS2 through receptor-induced endocytosis. Flg22 has been also implicated in the activation of *Arabidopsis* MPK3, MPK4, and MPK6 (Bethke et al. 2009). Two analogous MAPK signalling cascades are activated via receptor-like kinase complex of FLS2 and BAK1 (BRI1-associated kinase1) which perceives the signal of flg22 occurs at the plasma membrane (Yang et al. 2011). The primary MAPK cascade functioning downstream of the FLS2–BAK1 receptor network was recognized via transitory expression in protoplasts simultaneously with biochemical and genetic advances (Cristina et al. 2010). Xu and Zhang (2015) reported the triggering of MPK3 and MPK6 via MKK4 and MKK5 directing to the phosphorylation-dependent elicitation of the transcription factors WRKY22 and WRKY29 and FRK1 (flg22-induced receptor kinase 1) and early flg22 stimulated expression of genes, for instance, WRKY29, FRK1, and GST1. Camalexin biosynthesis functioning with upstream of PAD2 (phytoalexin deficient 2) and PAD3 also involves MAPK cascade including MPK3 and MPK6 (Frerigmann et al. 2015). The association among MAPK signalling cascades imparts tolerance against various pathogens. Moreover it has been reported that flg22 triggered MPK6 phosphorylates and alleviates ACC (1-amino-cyclopropane-1-carboxylic acid) synthase 6 (ACS6) involved in ethylene biosynthesis (Adachi et al. 2015). Though, MPK6 is not only implicated in PAMP-activated ethylene biosynthesis but also triggered through ethylene via MKK9 (Zhao and Guo 2011). Flg22 stimulated MPK3 which advanced further to activate VirE1-interacting protein 1 (VIP1), a transcription factor, that following phosphorylation is shifted from the cytoplasm to the nucleus to stimulate the expression of pathogenesis-associated genes, for example, PR1 (Šamajová et al. 2013a). Another MAPK module consists of MEKK1, MKK1/MKK2, and MPK4 that downregulates defense reaction (Tena et al. 2011). For such defense reaction, MEKK1 is requisite for particular triggering of MPK4 and excluding MPK3/MPK6 signalling pathway. Higher contents of SA in *mpk4* mutant advance to the genetic expression of pathogenesis-related genes and improved resistance against pathogen (Meng and Zhang 2013). Further, JA-induced expression of defensive proteins is requisite for MPK4 since triggering of particular genes was hampered in the *mpk4* mutant (Koornneef and Pieterse 2008). Conversely, stimulation of systemic acquired resistance in *mpk4*, *mekk1* or *mkk1*, *mkk2* double mutants generates particular dwarf phenotype which likely results from lesser cell dimensions (Gao et al. 2008). Such SA-mediated response relies on stimulation through WRKY transcription factors WRKY25 and WRKY33 via MKS1 which was established to interrelate with MPK4 (Andreasson et al. 2005). Bethke et al. (2012) reported the incidence of a nuclear MPK4–WRKY33 network in *Arabidopsis* indicated prior to pathogenic attack. MKS1 has demonstrated functional role in management and control of nuclear complex mks1 mutant. García and Parker (2009) reported elicitation with flagellin or *Pseudomonas syringae* trigger MPK4 and phosphorylate MSK1, leading to generation of MKS1 and WRKY33 from MPK4 and WRKY33 aiming the promoter of PAD3 encoding for enzyme requisite for camalexin biosynthesis. Mao et al. (2011) illustrated the functional role of MPK3 and MPK6 for the commencement of camalexin biosynthesis in *Arabidopsis* infested with fungal pathogen *Botrytis cinerea*. This occurs via phosphorylation of a pathogen-inducible transcription factor WRKY33 through MPK3/MPK6 which augments WRKY33 action in enhancing the downstream genetic expression of camalexin biosynthetic genes (Buscaill and Rivas 2014). Thus, MPK4 is requisite for stimulation of camalexin biosynthesis through bacterial pathogen (Qiu et al. 2008), except those implicated in camalexin stimulation by fungal pathogen (Mao et al. 2011). It can be speculated that MKK4/5–MPK3/6 and MKK1/2–MPK4 of MAPK cascade are triggered through flg22 function antagonistically. Doczi et al. (2012) evaluated interdependence between the MAPK modules as in *mkk1* mutant, flg22 mediated triggering of MPK4, MPK3 was hampered, and MPK6 is also rendered malfunctioned. Hence, the flg22 stimulated upregulation of defense reaction via MKK4/MKK5–MPK3/MPK6 pathway and downregulation through MKK1/MKK2–MPK4 pathway, signifying their role as potent entrants in innate immune response and resistance against pathogens (Mészáros et al. 2006). In addition to flg22, there are diverse fungal and bacterial PAMPs, for instance, chitin, harpin, elongation factor thermo-unstable (EF-Tu), and Nep1-like protein (NLP), that can stimulate MPK3/MPK6 or MPK4 and also upregulate pathogen-related genes (Vidhyasekaran 2014). Involvement of MAPK cascades in pathogen signal transduction is also well studied in tomato, tobacco, rice, parsley, and cotton. In tobacco, there are at least two MAPK pathways activated by inoculation with tobacco mosaic virus (TMV). The primary component is composed of MEK2—SIPK (salicylic acid-induced protein kinase) (Zhang and Liu 2001) and WIPK (wound-induced protein kinase) (Yap et al. 2005)—and leads to heavy metal tolerance. Another includes NPK1–MEK1–NTF6 and assuages defense response against TMV (Liu et al. 2004). Additionally, orthologs of SIPK and WIPK, for example, MPK1/2 and MPK3 in tomato, SIMK and SAMK in alfalfa, MAPK5 in rice, and MPK6 in parsley, also signify imperative function in defense-associated signal transduction (Zhang and Klessig 2001).

3.4 Plant Hormones Induced MAPK Activity

Nakagami et al. (2005) illustrated imperative role of MAPK cascades not only in defense against diverse biotic and abiotic stresses but also in hormone signalling in plants. Fujita et al. (2006) reported implication of plant hormones in cross talk among abiotic and biotic stress signalling pathways. Ethylene and jasmonic acid were rendered most significant for establishing the appropriate plant defense strategy against varied biotic and abiotic stresses (Cheong et al. 2002). However, triggering of MAPK cascade has been found requisite for the stimulation in biosynthesis of both these stress hormones. Gao et al. (2003) confirmed localization of ethylene receptor ETR1 (ethylene response 1) in endoplasmic-reticulum related with CTR1 (constitutive triple response 1) and a Raf-like MAP3K. Hahn and Harter (2009) reported ETR1 inhibited signal transduction pathway in absence of ethylene via downregulation of CTR1. Subsequently, dilapidation of the transcription factor EIN3 happened through 26S proteasome advancing to the obstruction of the downstream transcription cascade (An et al. 2010). Triggering of 47 kDa protein later recognized as MPK6 resulted following exposure to ethylene precursor ACC (Zhao and Guo 2011). Yoo et al. (2008) recognized new MKK9–MPK3/6 cascade that phosphorylates and stabilizes EIN3 through ethylene signalling. Cristina et al. (2010) reported inactivation of CTR1 pathway and activation of MKK9–MPK3/6 pathway as ACC was applied exogenously. MKK9 is transported to the nucleus from the cytoplasm where it activates nuclear MPK3 and MPK6 and phosphorylates and stabilizes EIN3 which then downstream transcription process (Yoo et al. 2008; Zhou et al. 2009). Yoo et al. (2008) reported in *Arabidopsis* mkk9 mutant plants exhibit malfunction manifested as ethylene insensitivity due to inactivation of MPK3 and MPK6. This entails simultaneous operation of up- and downregulating MAPK pathways which were incorporated into maintenance of EIN3 via phosphorylation, protein stabilization, and downstream transcription processes. Furthermore, MPK6 regulates the ethylene biosynthesis through phosphorylation and stabilization of enzymes ACS 2 and 6 (Liu and Zhang 2004). Takahashi et al. (2007) confirmed JA as significant contender in defense response against diverse stresses and developmental signal as it triggers MKK3–MPK6 cascade in *Arabidopsis*. The genetic evaluation through loss-of-function and gain-of function mutants of the MKK3–MPK6 cascade exhibited JA-stimulated cascade which downregulates AtMYC2, thereby influencing both JA-dependent gene expression and decline in root growth (Takahashi et al. 2007).

AP2C1, an MAPK phosphatase in *Arabidopsis*, is known to regulate MAPK functions and the JA reservoir. Schweighofer et al. (2007) reported dephosphorylation of MPK4 and MPK6 which leads to instant perception of wound activated signals through AP2C1 regulation. Taj et al. (2010) reported auxin-mediated MAPK signalling accompanied by NPK1 (a MAP3K) in tobacco and ANP1 (a NPK1 ortholog) in *Arabidopsis* protoplasts. Mockaitis and Howell (2000) reported instant activation of anonymous MAPKs in *Arabidopsis* root with diverse auxins. Dai et al. (2006) demonstrated MKK7 as downregulating factor in *Arabidopsis* for polar auxin transport as evaluated by genetic analysis. Lee et al. (2009) recognized MPK12 as downregulator of auxin signalling and as a substrate of MAPK phosphatase called IBR5 (indole-3-butyric acid response 5) in *Arabidopsis*. This MPK12 exclusively interrelates with IBR5 phosphatase, whereas phosphorylated MPK12 can be inactivated by phosphatase (Lee et al. 2009). Transgenic plants exhibiting lower genetic expression of the MPK12 display auxin sensitivity and adequate ABA sensitivity. Though, ibr5 mutant plants exhibited impaired responses to both auxin and ABA. Decline in MPK12 in an ibr5 milieu moderately liberated the ibr5 auxin-insensitive phenotype.

3.5 MAPK Modules Involved Both in Plant Development and in Stress Response

Extensive research demonstrated diverse stress-stimulated MAPKs and their positive regulators, for example, MAP2Ks and MAP3Ks are also involved in the maintenance of various plant developmental pathways. Kiegerl et al. (2000) demonstrated SIMK (stress-induced MAPK) from *Medicago sativa* in addition to its functional role in abiotic and biotic stress has also been implicated in the root hair formation and development via actin cytoskeleton association. This was the preliminary investigation combining genetic and cell biological expertise to reveal the significance of plant MAPK in the developmental process. Various allied MAPKs identified in alfalfa and tobacco were discreetly expressed throughout the cell cycle and expected to be associated with the maintenance of cell division. Savatin et al. (2014) reported MAP3Ks called ANP2/3 in *Arabidopsis* which was known to be involved in the last stage of cell division and cytokinesis. Takahashi et al. (2010) elaborated association between microtubular motor kinesins and microtubule bundling proteins of the MAP65 family suggesting the molecular interface of MAPKs with proteins. Furthermore, cytokinetic anp2/anp3 mutants also exhibited anomalous mitotic microtubules (Beck et al. 2011). Moreover, various MAPKs, for example, MPK18, MPK6, and MPK4, were expected to manage and/or correlate with cortical microtubules, consequently contributing in the establishing of plant cell shapes (Komis et al. 2011).

Stomata development characterizes an excellent example for developmentally regulated process governed via both stress-stimulated and developmentally activated MAPK cascades. In *Arabidopsis*, this section consists of YODA, MKK4/MKK5, and MPK3/MPK6 (Wang et al. 2007). Downstream mark of this pathway is a transcription factor SPEECHLESS (Lampard et al. 2008), whereas the upstream activator is generally a protein kinase known as SHORT SUSPENSOR (Bayer et al. 2009). Furthermore, it was anticipated in these research investigations that the same MAPK cascades can incorporate environmental and developmental signals to complete appropriate stomata growth and implementation (Lampard et al. 2008).

Bayer et al. (2009) reported an exceptionally equivalent MAPK-dependent signalling cascade controlling asymmetric cell division that functions throughout embryo development in *Arabidopsis*. Furthermore, MPK6 has been found involved in the control of cell division polarity throughout the post-embryogenic growth of seedling roots together with anther and inflorescence development (Müller et al. 2010). Finally, transcriptomic investigations demonstrated discreet control of stress-stimulated genes in MAP3K mutants anp2/3 (Krysan et al. 2002), yoda (Bergmann et al. 2004), and mekk1 along with mkk1/2 and mpk4 mutants (Pitzschke et al. 2009).

3.6 Strategies to Elucidate Stress-Stimulated MAPKs and Allied Plant Stress Tolerance

Several "omic" strategies prevail currently for the comprehensive elucidation of plant MAPKs and their specific purpose in stress tolerance, alongside transcriptomics, proteomics, and phosphoproteomics (exclusively the posttranslational modification of proteins). General genome transcriptomic databases are publically available for *Arabidopsis* and other crops such as rice, soybean, wheat, barley, maize, tomato, tobacco, or poplar. These databases comprise of extensive dataset of genes which illustrate modified expression under diverse stress and hormonal conditions and following chemical treatments at various stages competent of the definite growth zones or tissues of various organs. Identified individual genes can be clustered to hierarchical classes via MapMan program (http://gabi.rzpd.de/projects/MapMan) consistent with cellular processes and/or up- or downregulation of specific gene groups. Additionally, co-expression of genes in *Arabidopsis* and rice can be evaluated and related through ATTEDII web-based tool (http://atted.jp/) in order to approximate potential gene functions. Pitzschke et al. (2009) analyzed several signalling mutants, for example, *mekk1*, *mkk1/2*, and *mpk4* via transcriptomic strategy. Significant transcriptomic profiles of these signalling mutants can be also be associated, and proposition regarding recognized signalling cascades can be enhanced for those mutants which illustrate significant overlap in their transcriptomic datasets. Premise with reference to pathway association should be validated via more advanced biotechnological contrivances, for example, phenotyping, genetic crossing, bimolecular fluorescence complementation, and physiological investigations plus quantitative analysis of stress associated hormones. This was sufficiently established for MEKK1–MKK1/2–MPK4 pathway which functions as isolated project in plant innate immunity against fungal pathogen *Pseudomonas syringae* (Gao et al. 2008). Analogous mutants of this pathway are stunted, possess higher contents of salicylic acid (SA), and exhibit discreet expression of SA- and redox-responsive genes.

Programs like PLACE (www.dna.affrc.go.jp/PLACE/) or PlantCARE (http://sphinx.rug.ac.be:8080/PlantCARE/) can identify transcriptional regulatory regions in co-expressed genes. These programs recognize cis-elements within a set of diverse promoters. Consequently, motif plethora in a particular promoter can be assessed to the genomic setting frequency and the statistical involvement of the reinforcement of candidate motifs in promoters estimated via POBO contrivance (http://ekhidna.biocenter.helsinki.fi/poxo/pobo/pobo). Identified DNA motifs should be evaluated via overexpression of mock promoter assemblage and entrant transcription factors in transformed protoplasts/plants (Pitzschke et al. 2009). Commencement or suppression of genes enclosing candidate DNA target motifs recognized by transcription factors is approximated, as it was currently confirmed in the case of VIP1 transcription factor (Pitzschke et al. 2009). Transcription factors can be estimated as the ultimate phase indicating signalling cascade, and phylogenetic examination might afford for preliminary signal regarding their arrangement or approaching DNA target motifs. Some transcription factors such as WRKY demonstrate partiality for specific spacing among adjoining DNA motifs (W boxes) which is significant for their transcriptional activity (Bhat et al. 2013).

Regardless of certain possibility of entire genome transcriptomic progress, they have also limitations such as enhanced costs of experimentation, vital to control enormous datasets which necessitate bioinformatics investigation, and need of data regarding posttranscriptional alteration of gene products, particularly proteins which are frequently central for their validation. Signalling pathways and networks can be built via proteomic strategies. Data of interactive proteins can be identified via yeast two-hybrid (Y2H) screens or mass spectrometry examination of purified protein networks. Moreover, protein microchips can be utilized for selection of protein interface. Several softwares like in silico analysis (http://bar.utoronto.ca/interactions/cgi-bin/arabidopsis_interactions_viewer.cgi) present classy data regarding protein–protein interactions. Additionally, substantial with regard to kinase-induced signalling, phosphor–protein microarray chips were utilized for identification of putative MAPK contender substrate in *Arabidopsis* (Imamura et al. 2014). Organization of descriptive protein microarray-based proteomic system via threshold-based quantification permits recognition of 48 prospective substrates for MPK3 and 39 for MPK6. Extensive substrates were widespread for both kinases (Feilner et al. 2005). Furthermore, several novel signalling units surrounding varied MPKK/MPK pairs and 570 phosphorylated substrates of these units counting numerous WRKY and TGA transcription factors were identified employing high-density *Arabidopsis* microarrays enclose 2158 proteins (Šamajová et al. 2013b).

In this context, Y2H or protein microarray predicted association does not obviously signify that two proteins actually connect in planta, and contributing interrelating proteins must be scrutinized by advanced variety pattern, including their spatiotemporal expression pattern and their subcellular localization (Boruc et al. 2010). The inclusion of transcriptomic and proteomic information evidently assists the identification of superlative entrant genes and proteins involved in transduction of varied stress signals (Balderas-Hernández et al. 2013). Few of these applicants may be stress-responsive genes encoding for proteins regulated via MAPK-dependent phosphorylation. Phosphopeptide motif attribute for MAPKs can be recognized by comprehensive investigation of obtainable phosphopeptide sequences. It is desirable to confirm candidate phosphorylation motifs of individual proteins via in vitro and/or in vivo phosphorylation assays (Kersten et al. 2009). Moreover, a web-based Patmatch – TAIR can be used to screen for all *Arabidopsis* proteins anchoraging the same peptide motif (Zybailov et al. 2009). In précis, the expediency, vigor, and restrictions of applying a variety of transcriptomics and proteomics-based technology for decoding signalling pathways is still in its infancy (Choudhary and Mann 2010). Obviously, their exactly vast cipher of intricate merge harbors elevated prospective to extensively accelerate the advancement in signalling research by allocating experimentation to be intended in an extremely inundated approach and reinstate counter effort to an enormous level.

3.7 Stratagem for Genetic Manipulations of Kinases and Their Targets with Biotechnological Prospective

MAPK signalling cascades are activated, and they might modulate transcription via regulation of transcription factors; candidate genes and aiming proteins of these pathways may be distinguished via instantaneous implementation in transcriptomic, proteomic, and phosphoproteomic datasets of early stress reactions in varied crop species (Yang et al. 2013). Effectual description and genetic strategy of recognized objectives may be utilized not only for fundamental science and plant species, for instance, *Arabidopsis*, but recognized functional homologs will be of immense price also for enhancement of desirable traits, e.g., various stress resistance in crops (Pirie et al. 2013). Promoters stimulated via chemicals and/or dynamic only in definite plant tissues (tissue-specific promoters) may be employed to surmount adverse effects on plant growth and development leading from constitutive overexpression of genes (Molla et al. 2013). This approach was utilized to genetically influence VIP1 transcription factor controlling stress associated gene, e.g., PR1 (Bigeard et al. 2015). Toxic constitutive overexpression organization of VIP1 was substituted by expression from estradiol stimulated promoter which allocated efficient localization investigations of YFP-tagged VIP1 exhibiting rearrangement to the nucleus leading *Agrobacteria* and flg22 elicitation together with stimulation of PR1 gene (Šamajová et al. 2013b).

An alternative progress to aiming transcription factors is the identification of the upstream protein kinases themselves that are intervening the stress signals and eventually control the action of functional transcription factors. Data on the activation method of the protein kinases is, though, necessary for an accomplishment in this approach. For instance, the MKK2 kinase is implicated in cold and salt stress, and *mkk2* plants are hypersensitive to such stresses (Teige et al. 2004). Easy overexpression of the wild-type MKK2 gene has no advantageous consequence on stress tolerance. Conversely, the substitute of the threonine and serine residues of MKK2 that are usually phosphorylated via upstream regulation of MEKK1 by acidic amino acids capitulates an autoactive protein kinase. Transgenic plants that express constitutively active MKK2 are phenotypically usual but are currently extremely stress resistant (Teige et al. 2004).

To comprehend created data from replica plants, for example, *Arabidopsis* is certainly a principal attempt in agriculture. The main phase in this procedure involves the identification of the similar aspects in the crop plants. Although this predicament resembles easy task, genome development of extensive gene families can create this enterprise quite complex. An accommodating implement for these strategies is a bioinformatic machine at http://bioinfoserver.rsbs.anu.edu.au/utils/affytrees/, which offers data regarding the homologs of a protein of implication in other plant species. Although, there also exists of other performance, e.g., TILLING or oligonucleotide escorted mutagenesis to accomplish crops with a gene of implication enclosing an altered amino acid sequence. Ultimately, hybrid/artificial kinases can be contrived that adjust proteins other than their precise objectives or that confirm phosphorylation of a protein by outcompeting the precise altering upstream kinase. Principally those phosphorylation measures are widespread characteristic in the signalling of approximately all reactions and biological methods; this advance has elevated prospective for synthetic biology but also crop improvement.

3.7.1 Stress Tolerance in Arabidopsis with Genetically Modified MAPKs

Plants exploit two defensive mechanisms that assist them to resourcefully survive under diverse stress environments. Biotic stress arbitrated via pathogen originated compounds is apparent by transmembrane PRRs plus receptor proteins known as resistance proteins, involved in instant defense strategy. Although both biotic and abiotic stress responses frequently distribute parallel signalling components, appropriate understanding of abiotic stress-involved responses is delayed by intricacy of these processes which are inadequate to MAPK signalling pathways. However, the signalling pathway symbolized by MEKK1–MKK1/2–MPK4 is the backbone of pathogen-induced responses and function in mediating homeostasis of ROS which is central for maintaining biotic and abiotic stress tolerance (Pitzschke et al. 2009; Pitzschke and Hirt 2010). MPK4 is a functional regulator of plant defense system mostly on downregulation of SA signalling.

MPK4 is implicated in the regulation, stress signalling, and functional role in plant innate immunity. Besides, MPK3 and MPK6, the MPK4 was established third MAPK predisposed to the stimulation via hypoosmotic stress (Droillard et al. 2004). Furthermore, probable implication of MPK4 in hyperosmotic stress tolerance was also conjectured. Hyperosmotic stress responses in *mpk4* expected that MPK4 functions as negative regulator in addition to its significance in the downregulation in FLS2–MEKK1–MKK1/MKK2–MPK4 signalling pathway (Droillard et al. 2004). The similar pathway also symbolized significant regulatory system in the homeostasis of ROS. Transcriptomic analyses of *mekk1*, *mkk1/2*, and *mpk4* have displayed a complex of ROS-dependent genes and validated the function of this MAPK cascade as an incorporating constituent in ROS- and SA-initiated stress pathways (Pitzschke et al. 2009).

Furthermore, numerous members of MAPK components are implicated in abiotic stress responses and are essential for imparting tolerance to varied stress, e.g., salt, drought, or cold. Overexpression of MKK2 that targeted both MPK4 and MPK6 resulted in constitutive upregulation of numerous stress-induced genes, and the plants demonstrated elevated freezing and salt tolerance (Teige et al. 2004). Similar signalling method also seems to regulate hormone contents in response against pathogens, for example, *Pseudomonas* sp., and aids to sustain resistance against numerous bacterial pathogens (Kazan and Lyons 2014). MPK3 and MPK6 are significant in the control of physiological process stomatal opening/closure. Furthermore, these two kinases mutually with their upstream activators MKK4 and MKK5 are functional regulators of stomatal development and pattern (Wang et al. 2007). They function in definite collaboration with hydrogen peroxide and abscisic acid (ABA), and mutually they manage stomatal movements (An et al. 2008). Moreover, MPK3-related control of stomatal movement characterizes significant defense mechanism, which is able to efficiently avert bacterial invasion via stomata. These effects present obvious evidence of defensive approaches in biotic and abiotic stress frequently and based on comparable signalling methods.

MEKK1 is stimulated via touch, wounding, salinity, drought, or UV light abiotic stresses (Ara and Sinha 2014), and it triggers downstream signalling elements MKK1, MKK2, and MPK4 (Cristina et al. 2010). MEKK1–MPK4 pathway is functional regulator in ROS metabolism and signalling (Pitzschke et al. 2009). Conversely, similar to flagellin-stimulated pathogen reactions, homeostasis of ROS is also complex development with several signalling elements functioning independently. MPK3 and MPK6 are contributed in ROS-stimulated signalling as MPK3/MPK6 downregulations via RNAi technology formed plants hypersensitive to ozone, and triggering of the kinase in the single knockdown was considerably damaged (Wrzaczek et al. 2013). The association of ROS signalling pathways emerges to be still more intriguing with other regulating system adding up to its complexity. Yang et al. (2003) reported H_2O_2-mediated triggering of ROS biosynthesis in *Arabidopsis* through kinase of the NDP family AtNDPK2. Furthermore, NDPK2 exclusively interconnects with MPK3 and MPK6 (Jalmi and Sinha 2015). This disengages intimidation that NDPK2 might impart an upstream aspect involved in augmentation of MPK3/MPK6 plant defense pathways imparting the stress tolerance.

Furthermore, signalling pathways involved in the ROS homeostasis in plants have acquired hitherto an extra defensive strategy facilitating them to dispose of the hazardous types of molecular oxygen formed mainly from varied abiotic and biotic stresses. Scavenger enzymes, for example, superoxide dismutase, catalase, ascorbate peroxidase, or glutathione reductase, are controlled via ABA and provoked via H_2O_2 application (Danon 2012). Moreover, cross-talk flanking in Ca^{2+} and ROS signalling was suggested to incite antioxidant enzymes (Mazars et al. 2010). This is in accordance with further observation that Ca^{2+} channel obstructer hampered H_2O_2/ozone triggering of SIPK (orthologous to *Arabidopsis* MPK3 and MPK6) indicating that this upstream incident was mandatory in the ROS-stimulated signalling pathway (Lee and Ellis 2007). Amusingly, *Arabidopsis* catalase CAT1 was reported to be associated with MKK1–MPK6 signalling pathway (Xing et al. 2008), whereas both *mkk1* and *mpk6* mutants were capable to produce CAT1 following ABA application. Conversely, overexpression of these two signalling cascades led to enhanced ABA-dependent expression of CAT1 and generation of H_2O_2.

3.7.2 Tolerance Strategy in Plants Exhibiting Genetically Tailored MAPKs

Besides *Arabidopsis* pathogen-stimulated immune defense have been systematically illustrated in several crops. Research on tobacco MAPKs, WIPK, and SIPK (orthologs of MPK3 and MPK6) expresses corroboration that these two kinases were implicated in the directive of systemic resistance in opposition to tobacco mosaic virus arbitrated by SA (Zhang and Klessig 2001). Both WIPK and SIPK are also inclined to assorted mechanical and osmotic stresses. Another tobacco MAPK, Ntf4, with similar resemblance to WIPK/SIPK has been demonstrated (Ren et al. 2006). The entire signalling cascade allocates a general upstream MAPKK, NtMEK2, and they are all exposed to elicitor application. It was revealed that both elicitor application and overexpression of Ntf4 led to restricted cell death in instant HR (Ren et al. 2006). HR was also reported in disease-resistant plants via tomato and tobacco as representations. Through Pto-AvrPto resistance method, the authors described that tomato MAPKKK, MAPKKKα, was requisite in resistance and HR response advancing to pathogen-independent cell death (del Pozo et al. 2004).

Wheat homologs of MPK3/MPK6, called TaMPK3 and TaMPK6, were investigated throughout attuned interface of wheat with necrotrophic fungal pathogen *Mycosphaerella graminicola* (Rudd et al. 2008). This signalling pathway illustrates to be diverse in various features among different crop species. In *Arabidopsis*, tobacco, and others, these two signalling pathways were triggered concurrently against pathogen elicitations. On the contrary, TaMPK3 alone was transcriptionally triggered prior pathogen-stimulated programed cell death (PCD) (Rudd et al. 2008). Protein contents of TaMPK6 were compressed at this infection stage, which can be elucidated via accurate or vague protein dilapidation of this kinase (Rudd et al. 2008). Zaidi et al. (2009) recognized TMKP1 in wheat phosphatase, and its inclusive alliance with TaMPK3 and TaMPK6 was documented. Expression profiling revealed that the enzyme is stimulated following salt and osmotic stress. Moreover, localization testing with entire signalling cascade suggested that the TMKP1 can regulate subcellular localization of its interconnected legions. ROS signalling and its involvement in diverse abiotic stress signalling is a significant feature in biotechnological objectives in cash crops. In crops, huge understanding has been remunerated to technology supervision tolerance to drought, oxidative, and cold stress. MAPK from tobacco NPK1, an ortholog of *Arabidopsis* MAPK, exhibited wide range of tolerance against varied stresses. This NPK1 takes off ROS signalling analogous to MAPK modules functioning in the process. The decreased constitutive expression of NPK1 in maize results in improved freezing tolerance in the transgenic plants (Shou et al. 2004). Besides, these transgenic plants also revealed enhanced drought tolerance, which can be linked to a recognized defensive mechanism preserving the photosynthetic apparatus from desiccation.

Several new signalling pathways are added in ROS homeostasis which supplements our further understanding about this intricate equilibrium. Zhang et al. (2011) has identified constituent of the group C of MAPKs, the GhMPK2 from cotton. Enhanced expression of GhMPK2 in tobacco distributed the plants' tolerant to biotic stress which was directed via positive regulation of numerous pathogen-linked genes. Besides, positive regulation of scavenger antioxidant enzymes in the transgenic plants led to improved oxidative stress tolerance. The antioxidant enzymes indicate main effect for the plant to instantly metabolize ROS and avoid oxidative damage. Functional role of MAPK pathway imparting antioxidant defense has been associated with scavenger enzymes even in case of nitric oxide (NO) stress (Zhang et al. 2006). It has been suggested that NO functions in association with ABA-induced H_2O_2 generation, and subsequently, NO signalling is involved in enhanced expression of antioxidant enzymes in ABA signalling (Zhang et al. 2007). In maize, ABA-stimulated generation of H_2O_2 induces two other MAPKs, ZmMPK3 and ZmMPK5 (Wu et al. 2011). ZmMPK3 is susceptible to diverse signalling modules, e.g., jasmonic acid or salicylic acid, and is also activated against varied abiotic stresses, e.g., wounding, cold, drought, salinity, or UV light (Wang et al. 2010). ZmHK5 is implicated in upregulation process consisting ABA-induced generation and ROS supplying NADPH oxidase genes ZmRbohA-D (Lin et al. 2009). Generation of NADPH oxidase is a biphasic mechanism that can be moderately regulated with MAPK blockers and H_2O_2 scavengers (Shin et al. 2008). ZmHK5 is implicated in the triggering of the secondary stage of biphasic stimulation of NADPH oxidase which consecutively controls H_2O_2 generation. MAPK cascades and stress tolerance against diverse environmental stress have also been comprehensively analyzed in rice (Xiong and Yang 2003). This research study presented prolonged incubation under low temperatures (12 °C) was harmful for rice plants and may subsequently lead to male sterility and varied growth arrest phenotypes. Therefore, involvement of several MAPKs in the system has been exhibited accompanied by MAPKK OsMEK1 and MAPK OsMAP1 (Wen et al. 2002). The two signalling modules actually interconnect on the protein level and subsequently consist part of signalling pathway. On the other hand, enhanced expression of OsMAPK5 in transgenic plants consequently led to the enhanced multiple-stress tolerance. MAPK signal transduction pathway confers data of the extracellular surroundings to the cytoplasm, frequently leading to altered genetic expression and in the plant development. Although, accurate alteration of MAPKs impart beneficial effects in the organisms as evaluated via enhanced tolerance in response to environmental stimuli. Therefore, MAPKs and also MAPK-linked phosphatases are final targets of genetic alterations. Besides, immediate rationale of some modules of MAPK both in stress signalling and in plant development is also illustrated. A reflective understanding regarding control of MAPK pathway, for instance, via incorporated omics (transcriptomics, proteomics, phosphoproteomics, metabolomics, cellomics) progresses using comprehensively available software equipments for in silico database evaluation, might assist extended novel strategy to enhance stress tolerance in plants.

References

Abeles FB, Morgan PW, Saltveit ME Jr (2012) Ethylene in plant biology. Academic, New York

Adachi H, Nakano T, Miyagawa N, Ishihama N, Yoshioka M, Katou Y, Yoshioka H (2015) WRKY transcription factors phosphorylated by MAPK regulate a plant immune NADPH oxidase in Nicotiana benthamiana. Plant Cell 27(9):2645–2663

Alaoui-Sosse B, Genet P, Vinit-Dunand F, Toussaint ML, Epron D, Badot PM (2004) Effect of copper on growth in cucumber plants (*Cucumis sativus*) and its relationships with carbohydrate accumulation and changes in ion contents. Plant Sci 166:1213–1218

An Z, Jing W, Liu Y, Zhang W (2008) Hydrogen peroxide generated by copper amine oxidase is involved in abscisic acid-induced stomatal closure in *Vicia faba*. J Exper Bot 59(4):815–825

An F, Zhao Q, Ji Y, Li W, Jiang Z, Yu X, Zhang S (2010) Ethylene-induced stabilization of ETHYLENE INSENSITIVE3 and EIN3-LIKE1 is mediated by proteasomal degradation of EIN3 binding F-box 1 and 2 that requires EIN2 in Arabidopsis. Plant Cell 22(7):2384–2401

Andreasson E, Jenkins T, Brodersen P, Thorgrimsen S, Petersen NH, Zhu S, Newman MA (2005) The MAP kinase substrate MKS1 is a regulator of plant defense responses. The EMBO J 24(14):2579–2589

Anjum NA, Ahmad I, Mohmood I, Pacheco M, Duarte AC, Pereira E, Prasad MNV (2012) Modulation of glutathione and its related enzymes in plants' responses to toxic metals and metalloids- a review. Environ Exper Bot 75:307–324

Antunes MS, Morey KJ, Smith JJ, Albrecht KD, Bowen TA, Zdunek JK, … , Medford JI (2011) Programmable ligand detection system in plants through a synthetic signal transduction pathway. PLoS One 6(1):16292

Ara H, Sinha A (2014) Conscientiousness of mitogen activated protein kinases in acquiring tolerance for abiotic stresses in plants. Proc Indian Nat Sci Acad 80(2):211–219

Arimura GI, Ozawa R, Maffei ME (2011) Recent advances in plant early signaling in response to herbivory. Intern J Mol Sci 12(6):3723–3739

Arteca RN, Arteca JM (2007) Heavy-metal-induced ethylene production in *Arabidopsis thaliana*. J Plant Physiol 164(11):1480–1488

Balderas-Hernández VE, Alvarado-Rodríguez M, Fraire-Velázquez S (2013) Conserved versatile master regulators in signalling pathways in response to stress in plants. AoB Plants 5:033

Barbosa JS, Cabral TM, Ferreira DN, Agnez-Lima LF, de Medeiros SB (2010) Genotoxicity assessment in aquatic environment impacted by the presence of heavy metals. Ecotox Environ Safety 73(3):320–325

Bartwal A, Mall R, Lohani P, Guru SK, Arora S (2013) Role of secondary metabolites and brassinosteroids in plant defense against environmental stresses. J Plant Growth Regul 32(1):216–232

Bayer M, Nawy T, Giglione C, Galli M, Meinnel T, Lukowitz W (2009) Paternal control of embryonic patterning in Arabidopsis thaliana. Science 323(5920):1485–1488

Beck M, Komis G, Ziemann A, Menzel D, Šamaj J (2011) Mitogen-activated protein kinase 4 is involved in the regulation of mitotic and cytokinetic microtubule transitions in *Arabidopsis thaliana*. New Phytol 189(4):1069–1083

Bergmann DC, Lukowitz W, Somerville CR (2004) Stomatal development and pattern controlled by a MAPKK kinase. Science 304(5676):1494–1497

Bertin G, Averbeck D (2006) Cadmium: cellular effects, modifications of biomolecules, modulation of DNA repair and genotoxic consequences (a review). Biochimie 88(11):1549–1559

Bethke G, Unthan T, Uhrig JF, Pöschl Y, Gust AA, Scheel D, Lee J (2009) Flg22 regulates the release of an ethylene response factor substrate from MAP kinase 6 in Arabidopsis thaliana via ethylene signaling. Proc Nat Acad Sci 106(19):8067–8072

Bethke G, Pecher P, Eschen-Lippold L, Tsuda K, Katagiri F, Glazebrook J, … , Lee J (2012) Activation of the Arabidopsis thaliana mitogen-activated protein kinase MPK11 by the flagellin-derived elicitor peptide, flg22. Mol Plant-Microbe Interact 25(4):471–480

Bhat WW, Dhar N, Razdan S, Rana S, Mehra R, Nargotra A, … , Lattoo SK (2013) Molecular characterization of UGT94F2 and UGT86C4, two glycosyltransferases from Picrorhiza kurrooa: comparative structural insight and evaluation of substrate recognition. PloS one 8(9):e73804

Bigeard J, Colcombet J, Hirt H (2015) Signaling mechanisms in pattern-triggered immunity (PTI). Mol Plant 8(4):521–539

Binder BM, Chang C, Schaller GE (2012) Perception of ethylene by plants: ethylene receptors. Ann Plant Rev Plant Hormone Ethylene 44:117–145

Boller T, Felix G (2009) A renaissance of elicitors: perception of microbe-associated molecular patterns and danger signals by pattern-recognition receptors. Annu Rev Plant Biol 60:379–406

Boller T, He SY (2009) Innate immunity in plants: an arms race between pattern recognition receptors in plants and effectors in microbial pathogens. Science (New York, NY) 324(5928):742

Bolouri Moghaddam MR, Vilcinskas A, Rahnamaeian M (2015) Cooperative interaction of antimicrobial peptides with the interrelated immune pathways in plants. Mol plant pathol 17(3):464–71

Boruc J, Van den Daele H, Hollunder J, Rombauts S, Mylle E, Hilson P, … , Russinova E (2010) Functional modules in the Arabidopsis core cell cycle binary protein–protein interaction network. Plant Cell 22(4):1264–1280

Buscaill P, Rivas S (2014) Transcriptional control of plant defence responses. Curr Opin Plant Biol 20:35–46

Cabello JV, Lodeyro AF, Zurbriggen MD (2014) Novel perspectives for the engineering of abiotic stress tolerance in plants. Curr Opin Biotechnol 26:62–70

Cao S, Zheng Y, Wang K, Jin P, Rui H (2009) Methyl jasmonate reduces chilling injury and enhances antioxidant enzyme activity in postharvest loquat fruit. Food Chem 115(4):1458–1463

Cerana R, Malerba M (2015) Role of nitric oxide in heavy metal stress. In: Nitric oxide action in abiotic stress responses in plants, pp 181–192. Springer International Publishing, Switzerland

Cheong YH, Chang HS, Gupta R, Wang X, Zhu T, Luan S (2002) Transcriptional profiling reveals novel interactions between wounding, pathogen, abiotic stress, and hormonal responses in Arabidopsis. Plant Physiol 129(2):661–677

Choudhary C, Mann M (2010) Decoding signalling networks by mass spectrometry-based proteomics. Nat Rev Mol Cell Biol 11(6):427–439

Circu ML, Aw TY (2010) Reactive oxygen species, cellular redox systems, and apoptosis. Free Radic Biol Med 48(6):749–762

Cristina MS, Petersen M, Mundy J (2010) Mitogen-activated protein kinase signaling in plants. Annu Rev Plant Biol 61:621–649

Cuypers A, Karen S, Jos R, Kelly O, Els K, Tony R, … , Jaco V (2011) The cellular redox state as a modulator in cadmium and copper responses in Arabidopsis thaliana seedlings. J plant physiol 168(4):309–316

Cuypers A, Keunen E, Bohler S, Jozefczak M, Opdenakker K, Gielen H, … Remans T (2012) Cadmium and copper stress induce a cellular oxidative challenge leading to damage versus signalling. In: Metal toxicity in plants: perception, signaling and remediation. Springer, Berlin/Heidelberg, pp 65–90

Dai Y, Wang H, Li B, Huang J, Liu X, Zhou Y, … , Li J (2006) Increased expression of MAP KINASE KINASE7 causes deficiency in polar auxin transport and leads to plant architectural abnormality in Arabidopsis. Plant Cell 18(2):308–320

DalCorso G, Farinati S, Furini A (2010) Regulatory networks of cadmium stress in plants. Plant Signal Behav 5(6):663–667

Danon A (2012) Environmentally-induced oxidative stress and its signaling. In: Photosynthesis. Springer, Dordrecht, 319–330

Dar TA, Uddin M, Khan MMA, Hakeem KR, Jaleel H (2015) Jasmonates counter plant stress: a review. Environ Exp Bot 115:49–57

del Pozo O, Pedley KF, Martin GB (2004) MAPKKKα is a positive regulator of cell death associated with both plant immunity and disease. EMBO J 23(15):3072–3082

Demirevska-Kepova K, Simova-Stoilova L, Stoyanova Z, Hölzer R, Feller U (2004) Biochemical changes in barley plants after excessive supply of copper and manganese. Environ Exp Bot 52(3):253–266

Dóczi R, Brader G, Pettkó-Szandtner A, Rajh I, Djamei A, Pitzschke A, … Hirt H (2007) The Arabidopsis mitogen-activated protein kinase kinase MKK3 is upstream of group C mitogen-activated protein kinases and participates in pathogen signaling. Plant Cell 19(10):3266–3279

Dóczi R, Ökrész L, Romero AE, Paccanaro A, Bögre L (2012) Exploring the evolutionary path of plant MAPK networks. Trends Plant Sci 17(9):518–525

Droillard MJ, Boudsocq M, Barbier-Brygoo H, Laurière C (2004) Involvement of MPK4 in osmotic stress response pathways in cell suspensions and plantlets of Arabidopsis thaliana: activation by hypoosmolarity and negative role in hyperosmolarity tolerance. FEBS Lett 574(1):42–48

Emamverdian A, Ding Y, Mokhberdoran F, Xie Y (2015) Heavy metal stress and some mechanisms of plant defense response. Sci World J 2015:1–18

Erb M, Meldau S, Howe GA (2012) Role of phytohormones in insect-specific plant reactions. Trends Plant Sci 17(5):250–259

Feilner T, Hultschig C, Lee J, Meyer S, Immink RG, Koenig A, … Cahill DJ (2005) High throughput identification of potential Arabidopsis mitogen-activated protein kinases substrates. Mol & Cell Proteomics 4(10):1558–1568

Fiala R, Repka V, Čiamporová M, Martinka M, Pavlovkin J (2015) Early cadmium-induced effects on reactive oxygen species production, cell viability and membrane electrical potential in grapevine roots. VITIS-J Grapevine Res 54(4):175–182

Frerigmann H, Glawischnig E, Gigolashvili T (2015) The role of MYB34, MYB51 and MYB122 in the regulation of camalexin biosynthesis in Arabidopsis thaliana. Front Plant Sci 6:654

Fujita M, Fujita Y, Noutoshi Y, Takahashi F, Narusaka Y, Yamaguchi-Shinozaki K, Shinozaki K (2006) Crosstalk between abiotic and biotic stress responses: a current view from the points of convergence in the stress signaling networks. Curr Opin Plant Biol 9(4):436–442

Gallego SM, Pena LB, Barcia RA, Azpilicueta CE, Iannone MF, Rosales EP, … Benavides MP (2012) Unravelling cadmium toxicity and tolerance in plants: insight into regulatory mechanisms. Environ Exp Bot 83:33–46

Gao Z, Chen YF, Randlett MD, Zhao XC, Findell JL, Kieber JJ, Schaller GE (2003) Localization of the Raf-like kinase CTR1 to the endoplasmic reticulum of Arabidopsis through participation in ethylene receptor signaling complexes. J Biol Chem 278(36):34725–34732

Gao M, Liu J, Bi D, Zhang Z, Cheng F, Chen S, Zhang Y (2008) MEKK1, MKK1/MKK2 and MPK4 function together in a mitogen-activated protein kinase cascade to regulate innate immunity in plants. Cell Res 18(12):1190–1198

García AV, Parker JE (2009) Heaven's Gate: nuclear accessibility and activities of plant immune regulators. Trends Plant Sci 14(9):479–487

Gill SS, Tuteja N (2010) Reactive oxygen species and antioxidant machinery in abiotic stress tolerance in crop plants. Plant Physiol Biochem 48(12):909–930

Gill SS, Tuteja N (2011) Cadmium stress tolerance in crop plants: probing the role of sulfur. Plant Signal Behav 6(2):215–222

Gill SS, Anjum NA, Hasanuzzaman M, Gill R, Trivedi DK, Ahmad I, … Tuteja N (2013) Glutathione and glutathione reductase: a boon in disguise for plant abiotic stress defense operations. Plant Physiol Biochem 70:204–212

Glick BR (2015) Stress control and ACC deaminase. In: Principles of plant-microbe interactions. Springer International Publishing, Cham, 257–264

Guzel S, Terzi R (2013) Exogenous hydrogen peroxide increases dry matter production, mineral content and level of osmotic solutes in young maize leaves and alleviates deleterious effects of copper stress. Bot Stud 54(26):2–10

Hahn A, Harter K (2009) Mitogen-activated protein kinase cascades and ethylene: signaling, biosynthesis, or both? Plant Physiol 149(3):1207–1210

Hanaka A, Wójcik M, Dresler S, Mroczek-Zdyrska M, Maksymiec W (2016) Does methyl jasmonate modify the oxidative stress response in Phaseolus coccineus treated with Cu? Ecotoxicol Environ Saf 124:480–488

Haribabu TE, Sudha PN (2011). Effect of heavy metals copper and cadmium exposure on the antioxidant properties of the plant Cleome gynandra. Int J Plant Animal Environ Sci 1(2). ISSN 2231–4490

Hédiji H, Djebali W, Belkadhi A, Cabasson C, Moing A, Rolin D, … Chaïbi W (2015) Impact of long-term cadmium exposure on mineral content of Solanum lycopersicum plants: consequences on fruit production. S Afr J Bot 97:176–181

Hernández LE, Ortega-Villasante C, Montero-Palmero MB, Escobar C, Carpena RO (2012) Heavy metal perception in a microscale environment: a model system using high doses of pollutants. In: von Gupta DK, Luisa MS (eds) Metal toxicity in plants: perception, signaling and remediation. Springer, Berlin/Heidelberg, pp 23–39

Hu X, Neill S, Cai W, Tang Z (2003) Hydrogen peroxide and jasmonic acid mediate oligogalacturonic acid-induced saponin accumulation in suspension-cultured cells of Panax ginseng. Physiol Plant 118(3):414–421

Huang GT, Ma SL, Bai LP, Zhang L, Ma H, Jia P, … Guo ZF (2012) Signal transduction during cold, salt, and drought stresses in plants. Mol biol rep 39(2):969–987

Imamura H, Sugiyama N, Wakabayashi M, Ishihama Y (2014) Large-scale identification of phosphorylation sites for profiling protein kinase selectivity. J Proteome Res 13(7):3410–3419

Islam E, Khan MT, Irem S (2015) Biochemical mechanisms of signaling: perspectives in plants under arsenic stress. Ecotoxicol Environ Saf 114:126–133

Jakubowska D, Janicka-Russak M, Kabała K, Migocka M, Reda M (2015) Modification of plasma membrane NADPH oxidase activity in cucumber seedling roots in response to cadmium stress. Plant Sci 234:50–59

Jalmi SK, Sinha AK (2015) ROS mediated MAPK signaling in abiotic and biotic stress-striking similarities and differences. Front Plant Sci 6:769

Jaspers P, Kangasjärvi J (2010) Reactive oxygen species in abiotic stress signaling. Physiol Plant 138(4):405–413

Jonak C, Nakagami H, Hirt H (2004) Heavy metal stress. Activation of distinct mitogen-activated protein kinase pathways by copper and cadmium. Plant Physiol 136(2):3276–3283

Jozefczak M, Remans T, Vangronsveld J, Cuypers A (2012) Glutathione is a key player in metal-induced oxidative stress defenses. Int J Mol Sci 13(3):3145–3175

Karuppanapandian T, Moon JC, Kim C, Manoharan K, Kim W (2011) Reactive oxygen species in plants: their generation, signal transduction, and scavenging mechanisms. Aust J Crop Sci 5(6):709

Kazan K (2015) Diverse roles of jasmonates and ethylene in abiotic stress tolerance. Trends Plant Sci 20(4):219–229

Kazan K, Lyons R (2014) Intervention of phytohormone pathways by pathogen effectors. Plant Cell 26(6):2285–2309

Kersten B, Agrawal GK, Durek P, Neigenfind J, Schulze W, Walther D, Rakwal R (2009) Plant phosphoproteomics: an update. Proteomics 9(4):964–988

Khan MN, Mohammad F, Mobin M, Saqib MA (2014) Tolerance of plants to abiotic stress: a role of nitric oxide and calcium. In: von Khan MN (ed) Nitric oxide in plants: metabolism and role in stress physiology. Springer International Publishing, Cham, pp 225–242

Kiegerl S, Cardinale F, Siligan C, Gross A, Baudouin E, Liwosz A, … Meskiene I (2000) SIMKK, a mitogen-activated protein kinase (MAPK) kinase, is a specific activator of the salt stress–induced MAPK, SIMK. Plant Cell 12(11):2247–2258

Komis G, Illés P, Beck M, Šamaj J (2011) Microtubules and mitogen-activated protein kinase signalling. Curr Opin Plant Biol 14(6):650–657

Koornneef A, Pieterse CM (2008) Cross talk in defense signaling. Plant Physiol 146(3):839–844

Krysan PJ, Jester PJ, Gottwald JR, Sussman MR (2002) An Arabidopsis mitogen-activated protein kinase kinase kinase gene family encodes essential positive regulators of cytokinesis. Plant Cell 14(5):1109–1120

Kumar K, Wankhede DP, Sinha AK (2013) Signal convergence through the lenses of MAP kinases: paradigms of stress and hormone signaling in plants. Front Biol 8(1):109–118

Lampard GR, MacAlister CA, Bergmann DC (2008) Arabidopsis stomatal initiation is controlled by MAPK-mediated regulation of the bHLH SPEECHLESS. Science 322(5904):1113–1116

Lee JS, Ellis BE (2007) Arabidopsis MAPK phosphatase 2 (MKP2) positively regulates oxidative stress tolerance and inactivates the MPK3 and MPK6 MAPKs. J Biol Chem 282(34):25020–25029

Lee JS, Wang S, Sritubtim S, Chen JG, Ellis BE (2009) Arabidopsis mitogen-activated protein kinase MPK12 interacts with the MAPK phosphatase IBR5 and regulates auxin signaling. Plant J 57(6):975–985

Lee Y, Kim YJ, Kim MH, Kwak JM (2016) MAPK cascades in guard cell signal transduction. Front Plant Sci 7:80

Lei L, Li Y, Wang Q, Xu J, Chen Y, Yang H, Ren D (2014) Activation of MKK9-MPK3/MPK6 enhances phosphate acquisition in Arabidopsis thaliana. New Phytol 203(4):1146–1160

Li X, Li C (2004) Is ethylene involved in regulation of root ferric reductase activity of dicotyledonous species under iron deficiency? Plant Soil 261(1–2):147–153

Lin F, Ding H, Wang J, Zhang H, Zhang A, Zhang Y, … JiangM (2009) Positive feedback regulation of maize NADPH oxidase by mitogen-activated protein kinase cascade in abscisic acid signalling. J Exp Bot erp157

Liu Y, Zhang S (2004) Phosphorylation of 1-aminocyclopropane-1-carboxylic acid synthase by MPK6, a stress-responsive mitogen-activated protein kinase, induces ethylene biosynthesis in Arabidopsis. Plant Cell 16(12):3386–3399

Liu Y, Schiff M, Dinesh-Kumar SP (2004) Involvement of MEK1 MAPKK, NTF6 MAPK, WRKY/MYB transcription factors, COI1 and CTR1 in N-mediated resistance to tobacco mosaic virus. Plant J 38(5):800–809

Liu XM, Kim KE, Kim KC, Nguyen XC, Han HJ, Jung MS, … , Chung WS (2010) Cadmium activates Arabidopsis MPK3 and MPK6 via accumulation of reactive oxygen species. Phytochemistry 71(5):614–618

Loake G, Grant M (2007) Salicylic acid in plant defence—the players and protagonists. Curr Opin Plant Biol 10(5):466–472

Lumbreras V, Vilela B, Irar S, Solé M, Capellades M, Valls M, Coca M (2010) MAPK phosphatase MKP2 mediates disease responses in Arabidopsis and functionally interacts with MPK3 and MPK6. Plant J 63(6):1017–1030

Maksymiec W, Krupa Z (2006) The effects of short-term exposition to Cd, excess Cu ions and jasmonate on oxidative stress appearing in Arabidopsis thaliana. Environ Exp Bot 57(1):187–194

Manikandan R, Sahi SV, Venkatachalam P (2015) Impact assessment of mercury accumulation and biochemical and molecular response of mentha arvensis: a potential hyperaccumulator plant. Sci World J 2015:715217

Mao G, Meng X, Liu Y, Zheng Z, Chen Z, Zhang S (2011) Phosphorylation of a WRKY transcription factor by two pathogen-responsive MAPKs drives phytoalexin biosynthesis in Arabidopsis. Plant Cell 23(4):1639–1653

Martin MH (2012) Biological monitoring of heavy metal pollution: land and air. Springer Science & Business Media, Dordrecht

Mazars C, Thuleau P, Lamotte O, Bourque S (2010) Cross-talk between ROS and calcium in regulation of nuclear activities. Mol Plant (Oxford University Press/USA) 3(4):706–718

Means AR (2013) Calmodulin: properties. intracellular localization. and multiple roles in cell regulation. Recent Prog Horm Res 37:333–367

Meng X, Zhang S (2013) MAPK cascades in plant disease resistance signaling. Annu Rev Phytopathol 51:245–266

Mészáros T, Helfer A, Hatzimasoura E, Magyar Z, Serazetdinova L, Rios G, … Bögre L (2006) The Arabidopsis MAP kinase kinase MKK1 participates in defence responses to the bacterial elicitor flagellin. Plant J 48(4):485–498

Mittra B, Ghosh P, Henry SL, Mishra J, Das TK, Ghosh S, … Mohanty P (2004) Novel mode of resistance to Fusarium infection by a mild dose pre-exposure of cadmium in wheat. Plant Physiol Biochem 42(10):781–787

Mockaitis K, Howell SH (2000) Auxin induces mitogenic activated protein kinase (MAPK) activation in roots of Arabidopsis seedlings. Plant J 24(6):785–796

Molla KA, Karmakar S, Chanda PK, Ghosh S, Sarkar SN, Datta SK, Datta K (2013) Rice oxalate oxidase gene driven by green tissue-specific promoter increases tolerance to sheath blight pathogen (Rhizoctonia solani) in transgenic rice. Mol Plant Pathol 14(9): 910–922

Mota R, Pereira SB, Meazzini M, Fernandes R et al (2015) Effects of heavy metals on Cyanothece sp. CCY 0110 growth, extracellular polymeric substances (EPS) production, ultrastructure and protein profiles. J Proteomics 120:75–94

Müller J, Beck M, Mettbach U, Komis G, Hause G, Menzel D, Šamaj J (2010) Arabidopsis MPK6 is involved in cell division plane control during early root development, and localizes to the pre-prophase band, phragmoplast, trans-Golgi network and plasma membrane. Plant J 61(2):234–248

Nakagami H, Pitzschke A, Hirt H (2005) Emerging MAP kinase pathways in plant stress signalling. Trends Plant Sci 10(7):339–346

Nakajima H, Yamamoto Y, Yoshitani A, Itoh K (2013) Effect of metal stress on photosynthetic pigments in the Cu-hyperaccumulating lichens Cladonia humilis and Stereocaulon japonicum growing in Cu-polluted sites in Japan. Ecotoxicol Environ Saf 97:154–159

Nazar R, Iqbal N, Masood A, Khan MIR, Syeed S, Khan NA (2012) Cadmium toxicity in plants and role of mineral nutrients in its alleviation. Am J Plant Sci 3:1476–1489

Okazaki Y, Saito K (2014) Roles of lipids as signaling molecules and mitigators during stress response in plants. Plant J 79(4):584–596

Oldroyd GE (2013) Speak, friend, and enter: signalling systems that promote beneficial symbiotic associations in plants. Nat Rev Microbiol 11(4):252–263

Opdenakker K, Remans T, Keunen E, Vangronsveld J, Cuypers A (2012) Exposure of Arabidopsis thaliana to Cd or Cu excess leads to oxidative stress mediated alterations in MAPKinase transcript levels. Environ Exp Bot 83:53–61

Osakabe Y, Yamaguchi-Shinozaki K, Shinozaki K, Tran LSP (2013) Sensing the environment: key roles of membrane-localized kinases in plant perception and response to abiotic stress. J Exp Bot 64(2):445–458

Ovečka M, Takáč T (2014) Managing heavy metal toxicity stress in plants: biological and biotechnological tools. Biotechnol Adv 32(1):73–86

Parmar P, Dave B, Sudhir A, Panchal K, Subramanian RB (2013) Physiological, biochemical and molecular response of plants against heavy metals stress. Int J Curr Res 5(1):80–89

Parween T, Jan S, Sharma MP, Mujib A, Fatma T (2011) Genotoxic impact of cadmium on root meristem of Vicia faba L. Russ Agric Sci 37(2):115–119

Pirie CM, De Mey M, Prather KLJ, Ajikumar PK (2013) Integrating the protein and metabolic engineering toolkits for next-generation chemical biosynthesis. ACS Chem Biol 8(4):662–672

Pitzschke A, Hirt H (2009) Disentangling the complexity of mitogen-activated protein kinases and reactive oxygen species signaling. Plant Physiol 149(2):606–615

Pitzschke A, Hirt H (2010) Bioinformatic and systems biology tools to generate testable models of signaling pathways and their targets. Plant Physiol 152(2):460–469

Pitzschke A, Djamei A, Bitton F, Hirt H (2009) A major role of the MEKK1–MKK1/2–MPK4 pathway in ROS signalling. Mol Plant 2(1):120–137

Poovaiah BW, Du L, Wang H, Yang T (2013) Recent advances in calcium/calmodulin-mediated signaling with an emphasis on plant-microbe interactions. Plant Physiol 163(2):531–542

Popova LP, Maslenkova LT, Yordanova RY, Ivanova AP, Krantev AP, Szalai G, Janda T (2009) Exogenous treatment with salicylic acid attenuates cadmium toxicity in pea seedlings. Plant Physiol Biochem 47(3):224–231

Popova LP, Maslenkova LT, Ivanova A, Stoinova Z (2012) Role of salicylic acid in alleviating heavy metal stress. In: von Ahmad P, Prasad MNV (eds) Environmental adaptations and stress tolerance of plants in the era of climate change. Springer, New York, pp 447–466

Poschenrieder C, Tolrà R, Barceló J (2006) Can metals defend plants against biotic stress? Trends Plant Sci 11(6):288–295

Posmyk MM, Kontek R, Janas KM (2009) Antioxidant enzymes activity and phenolic compounds content in red cabbage seedlings exposed to copper stress. Ecotoxicol Environ Saf 72(2):596–602

Potters G, Pasternak TP, Guisez Y, Palme KJ, Jansen MA (2007) Stress-induced morphogenic responses: growing out of trouble? Trends Plant Sci 12(3):98–105

Qi Y, Tsuda K, Glazebrook J, Katagiri F (2011) Physical association of pattern-triggered immunity (PTI) and effector-triggered immunity (ETI) immune receptors in Arabidopsis. Mol Plant Pathol 12(7):702–708

Qiu JL, Zhou L, Yun BW, Nielsen HB, Fiil BK, Petersen K, … Morris PC (2008) Arabidopsis mitogen-activated protein kinase kinases MKK1 and MKK2 have overlapping functions in defense signaling mediated by MEKK1, MPK4, and MKS1. Plant Physiol 148(1):212–222

Rejeb IB, Pastor V, Mauch-Mani B (2014) Plant responses to simultaneous biotic and abiotic stress: molecular mechanisms. Plants 3:458–475

Remans T, Opdenakker K, Smeets K, Mathijsen D, Vangronsveld J, Cuypers A (2010) Metal-specific and NADPH oxidase dependent changes in lipoxygenase and NADPH oxidase gene expression in Arabidopsis thaliana exposed to cadmium or excess copper. Funct Plant Biol 37(6):532–544

Ren D, Yang KY, Li GJ, Liu Y, Zhang S (2006) Activation of Ntf4, a tobacco mitogen-activated protein kinase, during plant defense response and its involvement in hypersensitive response-like cell death. Plant Physiol 141(4):1482–1493

Rudd JJ, Keon J, Hammond-Kosack KE (2008) The wheat mitogen-activated protein kinases TaMPK3 and TaMPK6 are differentially regulated at multiple levels during compatible disease interactions with Mycosphaerella graminicola. Plant Physiol 147(2):802–815

Šamajová O, Komis G, Šamaj J (2013a) Emerging topics in the cell biology of mitogen-activated protein kinases. Trends Plant Sci 18(3):140–148

Šamajová O, Plíhal O, Al-Yousif M, Hirt H, Šamaj J (2013b) Improvement of stress tolerance in plants by genetic manipulation of mitogen-activated protein kinases. Biotechnol Adv 31(1):118–128

Sandmann G, Boger P (1980) Copper-mediated lipid peroxidation processes in photosynthetic membranes. Plant Physiol 66:797–800

Savatin DV, Bisceglia NG, Marti L, Fabbri C, Cervone F, De Lorenzo G (2014) The Arabidopsis NPK1-related protein kinases ANPs are required for elicitor-induced oxidative burst and immunity. Plant Physiol 165(3):1188–1202

Schmitt FJ, Renger G, Friedrich T, Kreslavski VD, Zharmukhamedov SK, Los DA, … Allakhverdiev SI (2014) Reactive oxygen species: re-evaluation of generation, monitoring and role in stress-signaling in phototrophic organisms. Biochim Biophys -Bioenergetics 1837(6):835–848

Schweighofer A, Kazanaviciute V, Scheikl E, Teige M, Doczi R, Hirt H, … Buchala A (2007) The PP2C-type phosphatase AP2C1, which negatively regulates MPK4 and MPK6, modulates innate immunity, jasmonic acid, and ethylene levels in Arabidopsis. Plant Cell 19(7):2213–2224

Sharma SS, Dietz KJ (2009) The relationship between metal toxicity and cellular redox imbalance. Trends Plant Sci 14(1):43–50

Sharma P, Jha AB, Dubey RS, Pessarakli M (2012) Reactive oxygen species, oxidative damage, and antioxidative defense mechanism in plants under stressful conditions. J Bot 2012:1–26

Shin MH, Moon YJ, Seo JE, Lee Y, Kim KH, Chung JH (2008) Reactive oxygen species produced by NADPH oxidase, xanthine oxidase, and mitochondrial electron transport system mediate heat shock-induced MMP-1 and MMP-9 expression. Free Radic Biol Med 44(4):635–645

Shou H, Bordallo P, Fan JB, Yeakley JM, Bibikova M, Sheen J, Wang K (2004) Expression of an active tobacco mitogen-activated protein kinase kinase kinase enhances freezing tolerance in transgenic maize. Proc Nat Acad Sci USA 101(9):3298–3303

Shrivastava S, Shrivastav A, Sharma J (2015) Detoxification mechanisms of mercury toxicity in plants: a review. Recent Adv Biol Med 1:60

Singh R, Gautam N, Mishra A, Gupta R (2011) Heavy metals and living systems: an overview. Indian J Pharmacol 43(3):246

Sirhindi G, Sharma P, Singh A, Kaur H, Mir M (2015) Alteration in photosynthetic pigments, osmolytes and antioxidants in imparting copper stress tolerance by exogenous jasmonic acid treatment in Cajanus cajan. Int J Plant Physiol Biochem 7(3):30–39

Smékalová V, Doskočilová A, Komis G, Šamaj J (2014) Crosstalk between secondary messengers, hormones and MAPK modules during abiotic stress signalling in plants. Biotechnol Adv 32(1):2–11

Sobkowiak R, Deckert J (2004) The effect of cadmium on cell cycle control in suspension culture cells of soybean. Acta Physiologiae Plantarum 26(3):335–344

Solanki R, Dhankhar R (2011) Biochemical changes and adaptive strategies of plants under heavy metal stress. Biologia 66(2):195–204

Sörensson C, Lenman M, Veide-Vilg J, Schopper S, Ljungdahl T, Grøtli M, … Andreasson E (2012) Determination of primary sequence specificity of Arabidopsis MAPKs MPK3 and MPK6 leads to identification of new substrates. Biochem J 446(2):271–278

Steffens B (2014) The role of ethylene and ROS in salinity, heavy metal, and flooding responses in rice. Front Plant Sci 5:685

Sun M, Li YT, Liu Y, Lee SC, Wang L (2016) Transcriptome assembly and expression profiling of molecular responses to cadmium toxicity in hepatopancreas of the freshwater crab Sinopotamon henanense. Sci Rep 6:19405

Sunitha MS, Prashant S, Kumar A, Rao SRINATH, Narasu ML, Ravi Kishor PB (2013) Cellular and molecular mechanisms of heavy metal tolerance in plants: a brief overview of transgenic plants overexpressing phytochelatin synthase and metallothionein genes. *Plant Cell Biotechnol.* Mol Biol 14(1–2):33–48

Sytar O, Kumar A, Latowski D, Kuczynska P, Strzałka K, Prasad MNV (2013) Heavy metal-induced oxidative damage, defense reactions, and detoxification mechanisms in plants. Acta Physiol Plant 35(4):985–999

Szalai G, Kellős T, Galiba G, Kocsy G (2009) Glutathione as an antioxidant and regulatory molecule in plants under abiotic stress conditions. *Journal of.* Plant Growth Regul 28(1):66–80

Taj G, Agarwal P, Grant M, Kumar A (2010) MAPK machinery in plants: recognition and response to different stresses through multiple signal transduction pathways. Plant Signal Behav 5(11):1370–1378

Takac I, Schröder K, Zhang L, Lardy B, Anilkumar N, Lambeth JD, … Brandes RP (2011) The E-loop is involved in hydrogen peroxide formation by the NADPH oxidase Nox4. J Biol Chem 286(15):13304–13313

Takahashi F, Yoshida R, Ichimura K, Mizoguchi T, Seo S, Yonezawa M, … Shinozaki K (2007) The mitogen-activated protein kinase cascade MKK3–MPK6 is an important part of the jasmonate signal transduction pathway in Arabidopsis. Plant Cell 19(3):805–818

Takahashi Y, Soyano T, Kosetsu K, Sasabe M, Machida Y (2010) HINKEL kinesin, ANP MAPKKKs and MKK6/ANQ MAPKK, which phosphorylates and activates MPK4 MAPK, constitute a pathway that is required for cytokinesis in Arabidopsis thaliana. Plant Cell Physiol 51(10):1766–1776

Teige M, Scheikl E, Eulgem T, Dóczi R, Ichimura K, Shinozaki K, … Hirt H (2004) The MKK2 pathway mediates cold and salt stress signaling in Arabidopsis. Mol Cell 15(1):141–152

Tena G, Boudsocq M, Sheen J (2011) Protein kinase signaling networks in plant innate immunity. Curr Opin Plant Biol 14(5):519–529

Thao NP, Khan MIR, Thu NBA, Hoang XLT, Asgher M, Khan NA, Tran LSP (2015) Role of ethylene and its cross talk with other signaling molecules in plant responses to heavy metal stress. Plant Physiol 169(1):73–84

Thapa G, Sadhukhan A, Panda SK, Sahoo L (2012) Molecular mechanistic model of plant heavy metal tolerance. Biometals 25(3):489–505

Vandenbossche M, Jimenez M, Casetta M, Traisnel M (2014) Remediation of heavy metals by biomolecules-a review. Crit Rev Environ Sci Technol 45(15):1644–1704

Vass I (2012) Molecular mechanisms of photodamage in the Photosystem II complex. Biochim Biophys Acta Bioenergetics 1817(1):209–217

Vidhyasekaran P (2014) PAMP signals in plant innate immunity. Signal Commun Plants. Springer Netherlands. doi:10.1007/978-94-007-7426-1

Walters M (2015) The plant innate immune system. J Endocytobio and Cell Res 26:8–12

Wang H, Ngwenyama N, Liu Y, Walker JC, Zhang S (2007) Stomatal development and patterning are regulated by environmentally responsive mitogen-activated protein kinases in Arabidopsis. Plant Cell 19(1):63–73

Wang P, Du Y, Li Y, Ren D, Song CP (2010) Hydrogen peroxide–mediated activation of MAP kinase 6 modulates nitric oxide biosynthesis and signal transduction in Arabidopsis. Plant Cell 22(9):2981–2998

Wasternack C (2014) Action of jasmonates in plant stress responses and development—applied aspects. Biotechnol Adv 32(1):31–39

Wasternack C, Hause B (2013) Jasmonates: biosynthesis, perception, signal transduction and action in plant stress response, growth and development. An update to the 2007 review in Annals of Botany. Ann Bot 111(6):1021–1058

Weckx JE, Clijsters HM (1996) Oxidative damage and defense mechanisms in primary leaves of Phaseolus vulgaris as a result of root assimilation of toxic amounts of copper. Physiol Plant 96(3):506–512

Wen J-Q, Oono K, Imai R (2002) Two novel mitogen-activated protein signaling components, OsMEK1 and OsMAP1, are involved in a moderate low-temperature signaling pathway in rice. Plant Physiol 129(4):1880–1891

Wrzaczek M, Brosché M, Kangasjärvi J (2013) ROS signaling loops—production, perception, regulation. Curr Opin Plant Biol 16(5):575–582

Wu T, Kong XP, Zong XJ, Li DP, Li DQ (2011) Expression analysis of five maize MAP kinase genes in response to various abiotic stresses and signal molecules. Mol Biol Rep 38(6):3967–3975

Xiao S, Dai L, Liu F, Wang Z, Peng W, Xie D (2004) COS1: an Arabidopsis coronatine insensitive1 suppressor essential for regulation of jasmonate-mediated plant defense and senescence. Plant Cell 16(5):1132–1142

Xie G, Sasaki K, Imai R, Xie D (2014) A redox-sensitive cysteine residue regulates the kinase activities of OsMPK3 and OsMPK6 in vitro. Plant Sci 227:69–75

Xing Y, Jia W, Zhang J (2008) AtMKK1 mediates ABA-induced CAT1 expression and H_2O_2 production via AtMPK6-coupled signaling in Arabidopsis. Plant J 54(3):440–451

Xiong L, Yang Y (2003) Disease resistance and abiotic stress tolerance in rice are inversely modulated by an abscisic acid–inducible mitogen-activated protein kinase. Plant Cell 15(3):745–759

Xu J, Zhang S (2015) Mitogen-activated protein kinase cascades in signaling plant growth and development. Trends Plant Sci 20(1):56–64

Yang KA, Moon H, Kim G, Lim CJ, Hong JC, Lim CO, Yun DJ (2003) NDP kinase 2 regulates expression of antioxidant genes in Arabidopsis. Proc Jpn Acad Ser B 79(3):86–91

Yang DH, Hettenhausen C, Baldwin IT, Wu J (2011) The multifaceted function of BAK1/SERK3: plant immunity to pathogens and responses 1 to insect herbivores. Plant Signal Behav 6(9):1322–1324

Yang SH, Sharrocks AD, Whitmarsh AJ (2013) MAP kinase signalling cascades and transcriptional regulation. Gene 513(1):1–13

Yap YK, Kodama Y, Waller F, Chung KM, Ueda H, Nakamura K, … Sano H (2005) Activation of a novel transcription factor through phosphorylation by WIPK, a wound-induced mitogen-activated protein kinase in tobacco plants. Plant Physiol 139(1):127–137

Yoo SD, Cho YH, Tena G, Xiong Y, Sheen J (2008) Dual control of nuclear EIN3 by bifurcate MAPK cascades in C2H4 signalling. Nature 451(7180):789–795

Zaidi A, Khan MS, Ahemad M, Oves M (2009) Plant growth promotion by phosphate solubilizing bacteria. Acta Microbiol Immunol Hung 56:263–284

Zhang S, Klessig DF (2001) MAPK cascades in plant defense signaling. Trends Plant Sci 6(11):520–527

Zhang S, Liu Y (2001) Activation of salicylic acid–induced protein kinase, a mitogen-activated protein kinase, induces multiple defense responses in tobacco. Plant Cell 13(8):1877–1889

Zhang J, Zhou JM (2010) Plant immunity triggered by microbial molecular signatures. Mol Plant 3(5):783–793

Zhang Y, Chai TY, Dong J, Zhao W, An CC, Chen ZL, Burkard G (2001) Cloning and expression analysis of the heavy-metal responsive gene PvSR2 from bean. Plant Sci 161(4):783–790

Zhang A, Jiang M, Zhang J, Tan M, Hu X (2006a) Mitogen-activated protein kinase is involved in abscisic acid-induced antioxidant defense and acts downstream of reactive oxygen species production in leaves of maize plants. Plant Physiol 141(2):475–487

Zhang T, Liu Y, Yang T, Zhang L, Xu S, Xue L, An L (2006b) Diverse signals converge at MAPK cascades in plant. Plant Physiol Biochem 44(5):274–283

Zhang A et al (2007) Nitric oxide induced by hydrogen peroxide mediates abscisic acid-induced activation of the mitogen-activated protein kinase cascade involved in antioxidant defense in maize leaves. New Phytol 175(1):36–50

Zhang H, Xia Y, Wang G, Shen Z (2008) Excess copper induces accumulation of hydrogen peroxide and increases lipid peroxidation and total activity of copper–zinc superoxide dismutase in roots of Elsholtzia haichowensis. Planta 227(2):465–475

Zhang WJ, Jiang FB, Ou JF (2011) Global pesticide consumption and pollution: with China as a focus. Proc Int Acad Ecol Environ Sci 1(2):125–144

Zhao Q, Guo HW (2011) Paradigms and paradox in the ethylene signaling pathway and interaction network. Mol Plant 4(4):626–634

Zhou C, Cai Z, Guo Y, Gan S (2009) An Arabidopsis mitogen-activated protein kinase cascade, MKK9-MPK6, plays a role in leaf senescence. Plant Physiol 150(1):167–177

Zhou J, Xia XJ, Zhou YH, Shi K, Chen Z, Yu JQ (2014) RBOH1-dependent H_2O_2 production and subsequent activation of MPK1/2 play an important role in acclimation-induced cross-tolerance in tomato. J Exp Bot 65(2):595–607

Zipfel C (2009) Early molecular events in PAMP-triggered immunity. Curr Opin Plant Biol 12(4):414–420

Zybailov B, Sun Q, van Wijk KJ (2009) Workflow for large scale detection and validation of peptide modifications by RPLC-LTQ-Orbitrap: application to the Arabidopsis thaliana leaf proteome and an online modified peptide library. Anal Chem 81(19): 8015–8024

Use of Mycorrhiza as Metal Tolerance Strategy in Plants

4

Contents

Abstract

Arbuscular mycorrhizal fungi occur in almost every soil ecosystem, as well as in contaminated soils. Arbuscular mycorrhizal fungi improve plant growth and development by increasing mineral uptake or enhancing resistance to biotic and abiotic stresses. Plasma membrane function may be rapidly affected by heavy metals as seen by an increased leakage from cells in the presence of high concentrations of metals, particularly of Cu. Arbuscular mycorrhizal fungi occurs in the almost every soil ecosystems, as well as in contaminated soils. The main attribute of AMF is to enhance nutritional status of their hosts by obtaining phosphates, micronutrients, and water from soil. The main characteristic feature of AMF is uptake of toxic metals via the fungal hyphae and their accumulation in plant tissues, and mostly heat shock proteins (HSPs) showed increased expression in response to the abiotic stress.

Keywords

AMF • Heavy metals • Heat shock proteins

S. Jan, J.A. Parray, *Approaches to Heavy Metal Tolerance in Plants*,
DOI 10.1007/978-981-10-1693-6_4

4.1 Introduction

Arbuscular mycorrhiza fungi (AMF) interactions with plants are significant due to their nutritional and non-nutritional benefits in natural and man-made ecosystems (Shirmohammadi et al. 2014). AMFs play a major role as biofertilizers, bioprotectants, and biodegraders for enhancing plant productivity by improving mineral nutrition or increasing resistance or tolerance to biotic and abiotic stresses (Gao et al. 2010; Sanjay and Kumar 2011). AMFs are ubiquitous root symbionts with bizarre genetic and physiological diversity and are important components of major ecosystems mainly for bioremediation (Hossain et al. 2012). (Schutzendubel and Polle 2002). There are several schools of thoughts regarding the mechanisms for toxic metal direction and allocation in plant roots (Hossain et al. 2012). Various hypotheses may be like:

(i) Some heavy metals bound to cell wall and deposited in the vacuoles of AMF.
(ii) Siderophores also help in metal sequestration in the soil or into root apoplasm.
(iii) The metals may bound to metallothioneins or PCs inside the fungal or plant cells.
(iv) The metal transporters at the tonoplast of both plants and fungi catalyze the transport of metals from cytoplasm.

Hossain et al. (2012) reported that some AMFs like *Glomus intraradices* colonize *Medicago truncatula* under Zn stress (ZnSO4, 100 μM), and a sharp transcriptional increase of a GSH-dependent GST gene was observed which indicates that effective fungal heavy metal tolerance mechanisms may be crucial part for the increased tolerance of toxic metals and their accumulation in mycorrhizal plants. There are some main considerations of AMF association for the benefit of plants as it increases their fitness under metal stress conditions. Meanwhile, a complete investigation pertaining to role of AMF fungi on GSH-dependent HM tolerance of plants or other abiotic stresses will be helpful for the effective fugal strain and also the plant fungal association for the bioremediation purposes (Gadd 2010). However, ectomycorrhizas that are characteristic features of trees and shrubs are to be considered as effective in ameliorating the effects of metal toxicity on the host plants (Gosling et al. 2010). But the mechanisms for increase in metal tolerance levels are still not clear perhaps due to metal and fungal species specificity (Liu et al. 2015). For example, ectomycorrhizal fungus *Paxillus involutus* accumulates Zn and accordingly reduces the Zn level in *Pinus sylvestris* tissues; however, another fungal species *Thelephora terrestris* retained little Zn, but the Zn content of the host plant was elevated (Irena and Ajit 2011). In another report, it was noted that both mycorrhizal species *Suillus bovinus* and *Thelephora terrestris* protected *P. sylvestris* against Cu toxicity; however, Cu retained varied significantly between two fungi (Hall 2002). Similarly Zarei et al. (2010) noted the significant variation between the capabilities of five ectomycorrhizal fungi to grow in culture augmented with range of nine different heavy metals. It is obvious that fungi perhaps used the same strategies for tolerance to heavy metals as employed by higher plants by binding the metals to extracellular materials or sequestration in the vacuolar compartment (Wu et al. 2010). Besides the role of ectomycorrhizas in metal tolerance by the plants, some exclusion processes have been also proposed for restriction metals into host plants like hyphal sheath metal absorption, reduced entrée to the apoplast, fungal exudates' chelation, etc. (Hassan et al. 2015), and these exclusion processes are variable between different plant and fungal interactions. The phytoremediation of toxic metals in the case of maize root colonization by arbuscular mycorrhiza ultimately depends on growth conditions, fungus, and the toxic metal (Amal and Hasnah 2012) like *Viola calaminaria* growing on contaminated soil supports the growth of both maize and alfalfa than a commonly used *Glomus* species (Satish 2013). Further most of the work done so far have been focused root and plant cell uptake of N, P, S, Fe, Ca, K, and Cl (Ebrahim et al. 2014). However, very scant information is available on mobilization, uptake, and transport of some other environmentally hazardous toxic metals, such as Pb, Cd, Cu, Zn, U, Sr, and Cs. This may perhaps due to ability to remain sorbed to solid soil particles and plants have to mobilize them for the phytoextraction, and this process of mobilization of soil-bound metal ions can be carried out via different pathways (Vamerali et al. 2010).

4.2 Root Cell Wall and Exudates

The metal-specific tolerance by roots is rather a difficult process to understand; however, various reports have mentioned the role of proteins for the metal accumulation. The root cell wall is in direct contact with toxic metals in any environment, and adsorption onto the cell wall or plasma membrane is least effected due to limited metal activity at the root surface. The heavy metal-tolerant *Silene vulgaris* ssp. humilis accumulated vast variety of toxic elements in the epidermal cell walls, and the reason for binding may be either by a protein or as silicates (Dorota et al. 2015). Similarly root exudates have a definite role to play for metal sequestration mainly by metal chelators or siderophores that increases the uptake of metals. An important study by Salt and Kramer (2000) mentions the role of Ni-chelating exudates in detoxification strategy in some Ni-hyperaccumulating plants, and it was seen that Ni-chelating histidine and citrate was accumulated in the root exudates of non-hyperaccumulating plants vis-à-vis reducing Ni pollution in contaminated soils. The range of exudates released from roots is wide so there is possibility of having a tolerance or detoxification of many other heavy metals as well. The most studied root secretions like organic acids and their role in detoxification of toxic metals particularly aluminum (Al) have been studied in various plant species viz., buckwheat secretes oxalic acid from the roots in response to Al stress and accumulates non-toxic Al–oxalate in the leaves and in wheat and maize; there is also evidence of such secretion mediated by Al stress (Ma et al. 2005).

4.3 Plasma Membrane

The plasma membrane of plants is considered as most prone to heavy metal toxicity. The high concentrations of metals affect the functioning of plasma membrane which is obvious by an increased leakage of metal ions from cells. For example, Zn caused increased K^+ efflux from excised roots of *Agrostis capillaris* (Liu et al. 2008). Besides other reports mentioned that the cell membrane damage was the primary cause of Cu toxicity in roots of various plants like *Silene vulgaris*, *Mimulus guttatus*, wheat, etc. (Verbruggen et al. 2009). The plasma membrane damage might be due to either oxidation or cross-linking of protein thiols, inhibition of H^+-ATPase proteins, or structural changes in membrane lipids (Hussain et al. 2004).

The heavy metal tolerance involves the protection of plasma membrane integrity against heavy metal toxicity. It is reported that the metal-tolerant plants do not appear to possess enhanced tolerance to free radicals or ROS but rather depend on improved mechanisms for metal homeostasis (Bussche and Soares 2011). Further the metal-specific membranes against different metal ions vis-à-vis protection of membranes against oxidation and membrane leakage varies between species (Mganga et al. 2011). Some reports observed the maintenance of plasma membrane integrity in the presence of toxic metals which enhanced membrane repair after cell damage (Salt and Kramer 2000) that involves heat shock proteins or metallothioneins.

Further, the cell membranes are attributed with the property of metal homeostasis by different mechanisms by either blocking the entry or through other selective efflux mechanisms. However, energy-dependent efflux of toxic metal ions is observed in bacteria (Banerjee and Chandel 2011). It is observed that the metabolic pathways for metal ion uptake mechanisms are greater than inducible efflux systems (Williams et al. 2000); however, the exclusion or reduced uptake mechanisms in limited in higher plants. The *Holcus lanatus* roots take up both phosphate and arsenate different systems. It was observed that a lesser rate of uptake for PO_4 and As anions was seen in arsenate-tolerant genotypes because of absence of the high-affinity metal uptake system (Jia et al. 2012). The arsenate tolerance in *H. lanatus* utilizes both adaptive suppression of the high-affinity transport system and phytochelatin (PC) production (Hassinen et al. 2011). Richau and Schat (2009) have also described the plasma membrane transporter in *N. tobacum* with Ni tolerance and Pb hypersensitivity.

The active efflux of metal ions is considered to be an effective alternative strategy, in plants, in comparison to bacteria where efflux pumping is the basis for toxic ion resistance systems (Wang et al. 2011). Some of the efflux pumping systems have been identified for various metal ions like Cu, Cd, Zn, Co, Ni, etc. (Hassan and Aarts 2011). Even though there is not any confirmation in plants regarding the direct role for plasma membrane, efflux transporters in metal tolerance and recently some research have bared that plants also possess some classes of metal transporters like heavy metal CPxATPases, the Nramps, and the CDF (cation diffusion facilitator) family and the ZIP family for metal uptake and homeostasis (Williams et al. 2000; Guerinot 2010).

4.4 Expression and Role of Heat Shock Proteins (HSPs)

HSPs are expressed in response to various stresses in plants besides other living organisms as well and are classified on the basis of molecular size and expression in response stress conditions like heavy metals (Lavania et al. 2015). Besides normal functioning ability, they show protein protective and repair property under stress conditions.

There are numerous and detailed reports of HSP expression under stress conditions like low-molecular mass HSPs (16–20 kDa) were expressed in rice under heavy-metal stress. As in rice (Lee et al. 2008), HSP17 were observed to express in roots of *Armeria maritima* plants under Cu stress (Sarkar et al. 2013).

The larger HSP (HSP70) was also seen to cause Cd stress in cell cultures of *L. peruvianum* (Neumann et al. 1994). Further the report considered the role of HSP70 in membrane protection against Cd toxicity. The expression of HSP70 also increased in *Enteromorpha intestinalis* under Cu stress (Guan et al. 2010). Thus, HSPs are considered for membrane metal or enhanced repair mechanisms (Shakeel et al. 2011); however, more advanced work needs to be carried out in this direction. The concentrations of toxic heavy metals like Pb, Zn, Cu, Cd, and As were found accumulated in arbuscular mycorrhizal (AM)-associated plants in contaminated soils (Hossain et al. 2012). Among the different families, Gramineae and Compositae generally grow in dominance on contaminated sites. The commonly plants found at these sites are *Chrysanthemum moritolium*, *Cynodon dactylon*, and *Miscanthus floridulus* (Shirmohammadi et al. 2014). However, AM fungal colonization was found in the roots of *C. dactylon* and *P. vittata* in mine sites (Hossain et al. 2012). Further the concentration of As were found maximum in upper parts than in roots of *P. vittata* with 73 % mycorrhizal colonization, and *C. dactylon* showed higher accumulation of As with 85 % mycorrhizal colonization in its shoots (Gadd 2010). Nowadays, plants in association AM fungi are used for phytoremediation in contaminated soils (Liu et al. 2015).

The plants growing under metal-contaminated sites are generally tolerant to toxic metals and can be categorized into three groups:

(i) Excluders: The metal concentrations in the shoots are kept up to a critical value at a low level across a wide range of external metal concentrations (Vamerali et al. 2010).
(ii) Accumulators: The metal level is concentrated in plant tissues from lower to higher soil concentrations (Fijalkowski et al. 2012).
(iii) Indicators: Internal metal concentrations in the plant imitate external levels, i.e., soil concentrations (Ashraf et al. 2013).

But the metal tolerance in plants depends on various biological, chemical, and other physiological adaptations in contaminated areas, and AMF association with plants may be fruitful by providing a metal excluder barrier and improving nutritional status of plants (Hossain et al. 2012).

Evolutionary adaptation of AMF-associated plants to the contaminated areas is to be selected on the basis of variability and habitability in tolerance and higher fitness of tolerant and non-tolerant individuals on polluted and unpolluted sites which depends upon their several strategies particularly chelation with mycorrhizal fungi (Shirmohammadi et al. 2014). Further it is mentioned that AM fungi were associated with numerous plant species growing in industrially polluted areas and also support the growth and survival of plants (Liu et al. 2015).

4.5 Mechanism of Arbuscular Mycorrhizal (AM) Fungi for Phytoremediation

AMF occurs in most soil ecosystems, as well as in contaminated soils. They improve nutritional status of the plants by phosphate solubilization and acquisition of other nutrients and water (Liu et al. 2015). The toxic or heavy metals can be transported to plants via fungal hyphae which may be either by direct heavy metal uptake and root-to-shoot transport, i.e., phytoextraction, or AMF immobilizing the metal ions in soil, i.e., phytostabilization. The effective mycorrhizal association for cleaning the contaminated soils depends on the combination of AMF and metal and also on soil characteristics (Gadd 2010). Glomeromycota is considered to be the most abundant in major soil ecosystems (Schutzendubel and Polle 2002) and interacts with the roots of more than 80 % of plants by enhancing the nutrient uptake (Shirmohammadi et al. 2014). Spores and presymbiotic hyphae are generally sensitive to toxic metals in the absence of plants (Singh 2012). The spores from polluted soils are found tolerant to higher heavy metal concentrations than spores from uncontaminated soils (Zarei et al. 2010) which may be due to phenotypic plasticity rather than genotypic changes in the spores (Bonanno and Giudice 2010). Furthermore, the effect of Zn, Pb, and Cd on presymbiotic i.e., spore germination and hyphal extension and symbiotic including extraradical mycelial growth and sporulation fungal association showed *Glomus intraradices* to be more tolerant to heavy metals than *Glomus etunicatum* (Rees et al. 2014), confirming that metal tolerance depends upon fungal genotypes. Further the synergistic or antagonistic interactions are observed under mixtures of heavy metals which may either increase or decrease the toxicity. The addition of Zn was found antagonistic to the toxicity of Pb or Cd on presymbiotic hyphal growth, while Pb and Cd acted synergistically (Miransari 2011). Nowadays, the plants are generally introduced into soil without the established symbioses because AM fungi occur in contaminated soil and are adapted accordingly considered as prospective tool for bioremediation (Yang et al. 2015). The detailed mechanism for heavy metal tolerance in plants with the aid of AMF association is depicted in Fig. 4.1. The mechanism of heavy metal tolerance by plants is described by two processes, i.e., phytostabilization and phytoextraction.

4.5.1 Phytostabilization

The immobilization of metals within the rhizosphere is carried out by their precipitation within the soil or by adsorption onto the root surface or uptake/accumulation within roots (Zeng et al. 2010). Further the phytostabilization prevents spreading of metals into the soil as well as their leakage due to erosion from the soil. The main aspect of phytostabilization strategies is the ability of metal-tolerant plants with extensive root systems for preventing wind/water erosion of toxic metals (Amal and Hasnah 2012). AM fungi contribute to the immobilization of heavy metals in the soil by chelative property of metals inside the fungus, secretory compounds, phosphate solubilization, and adsorption onto fungal cell walls (Hossain et al. 2012). The AMF are generally known to produce a glomalin, an insoluble glyco-

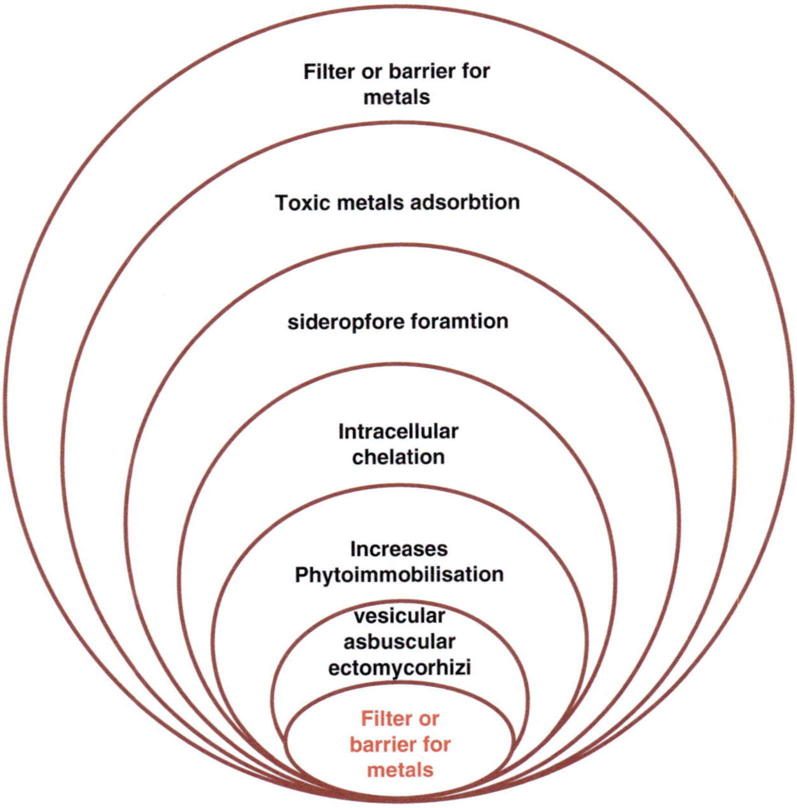

Fig. 4.1 Role of AMF in heavy metal tolerance by plants in forest ecosystem (Shirmohammadi et al. 2014)

protein and has a binding property of most toxic metals in the soil (Gonzalez-Chavez et al. 2004). Glomalin can be extracted from the soil together with bound HMs. Fungal strains with significant secretion of glomalin are considered more suitable for biostabilization efforts.

Further the binding of HMs to chitin reduces their concentration in the soil like the passive adsorption on the hyphae leads to binding of up to 0.5 mg Cd per mg dry biomass (Gamalero et al. 2003). This is due to the large surface area presented by fungi in the soil for binding of HM. The AMF tolerant-fungi posses four times more strong affinity for metals than roots in the soil (Joner et al. 2000). Further the Pb uptake and its immobilization were observed higher in mycorrhizal roots than non-mycorrhizal plants (Chaudhry et al. 2005). However it is observed that fungal vesicles as in plant and fungal vacuoles are involved in accumulating toxic metals which provides an additional detoxification mechanism (Elekes 2014). In one study, the *Glomus* isolate Br1 isolated from *Viola calaminaria* roots in metal-contaminated soil was allowed to colonize maize, alfalfa, barley, and *V. Calaminaria*; however, a common, not metal adapted, *G. intraradices* also allowed growth of plants, but to a lesser extent, and non-colonized plants didn't survive in the same contaminated soil (Oh et al. 2014). Further in another observation on maize, it was evident that toxic metals are relatively concentrated in inner parenchyma cells corroborating with AMF (Joner et al. 2000). The fungal hyphae is normally considered to have higher affinity to metals than plants maybe due to the fact that they became immobilized in the hyphal cell (Singh and Kalamdhad 2013). That represents strong metal detoxification mechanism. It is also associated with reduced gene expression of heavy metal inducible plant genes upon mycorrhizal colonization (Malik et al. 2010). Similarly, Cd is stabilized in the root system of AM-associated plants of pea, clover, and ribwort (Joschim et al. 2009), and this effect was assigned due to the changes in polyamine metabolism (Leung et al. 2007). It was observed that AMF association seems to create a more effective environment that allows roots to cope under high metal stress perhaps by accumulating metals at or in fungal structures (Hildebrandt et al. 2007). Further, there are also many reports about the role of fungus in protection to oxidative stress. For example, increased accumulation of Zn in the roots of AM plants like clover and maize has been reported, but under Zn limitation, mobilization of Zn and transfer to the shoot is enhanced by the AM symbiosis (Chen et al. 2005), determining the role of Zn as a micronutrient vis-à-vis its beneficial role for nutrient supply, which basically illustrates the mechanisms of AMF for immobilization of toxic/heavy metals in the soil or within roots for phytostabilization applications (Long et al. 2010). Besides the complex nature of system shall be noted about as each contaminated site has a specific sketch of toxicants for which a suitable fungal and plant combination must be established (Xun et al. 2015), and there are also other interactions in the soil that could either positively or negatively affect the AMF–plant association or detoxification process (Lin et al. 2012).

4.5.2 Phytoextraction

Phytoextraction is an advanced and effective technology for accumulation and transfer of heavy metals from root to shoots in contaminated soils (Cao et al. 2008; Salt and Kramer 2000). Besides phytoremediation these plants are harvested from the polluted sites and can be used for energy production called as phytomining (Jiang et al. 2010; Azevedo et al. 2012). However the major difficulty for rapid phytoremediation is bioavailability of metals in the soil, and overall efficiency depends on metal tolerance and biomass production by the plants. The property of hyperaccumulators is that they can enrich heavy metal by around 1000-fold as compared to non-accumulators that too enrich without showing toxicity (Mourato et al. 2012). It is observed that majority of plants produce very low concentration of chelates however the possible ways for accelerating the clean-up process even in non-hyperaccumulators include the addition of chelates or biological agents like AMF (Siddiqui et al. 2014). The most commonly used synthetic chelate is EDTA for Pb removal and accumulation in contaminated sites (Kramer et al. 2007). The addition of AMF is reported to further enhance the uptake and accumulation of heavy metals like As removal was seen in *P. vittata* (Leung et al. 2007). *Berkheya coddii* and *P. vittata* are recognized as a phytoextraction crops used mainly for phytomining of As, Ni, and other heavy metals (Salt et al. 1998). Further it is reported that the biomass of Ni hyperaccumulator was found higher in AM fungi colonized plants (Salt and Kramer 2000; Sanjay and Kumar 2011). A few reports regarding the phytoextraction like mycorrhizal tomato plants were found to have three times higher root and shoot biomass than non-colonized controls besides As and Ni accumulation and removal from soils (Liu et al. 2015). The enhanced biomass with higher Pi acquisition and Zn accumulation within the thyme shoots (*Thymus polytrichus*) has been documented (Wuana and Okieimen 2011). It has been reported that colonization by AM fungi to plants diminished the phytotoxic effect of higher Zn levels and increased mobilization of Zn from the soil (Park 2010).

Further the use of chelates can cause water pollution as metals may move into groundwater (Sanjay and Kumar 2011). In this perspective, plant colonization by AMF leads to increased uptake and metal accumulation in aboveground tissues; however, in some cases mycorrhizal colonization leads to accumulation in roots (Shirmohammadi et al. 2014). It is pertinent to mention that phytoextraction depends on the appropriate conditions, i.e., AMF–plant association for a given contaminated site and various factors are important for heavy metal detoxification by plants (Hossain et al. 2012):

- Various chelating agents are produced from both plants (organic acids) and AMFs (glomalin).
- Metal binding to cell wall of both plants and AMF.
- Plasma membrane acts as a living and selective barrier of toxic metals.
- Specific and nonspecific metal transporters and pores in the plasma membrane of plants and fungi (active and passive import).
- Chelates in the cytosol, e.g., metallothioneins (plants and fungi), organic acids, amino acids, and metal-specific chaperons (shown for plants, assumed for AM fungi).
- Export via specific or nonspecific active or passive transport from plant or fungal cells.
- Sequestration of HM in the vacuole of plant and fungal cells.
- Transport of HM in the hyphae of the fungus.
- In arbuscules, metal export from the fungus and import into plant cells via active or passive transport.

4.6 Developmental Patterns of AMF During Heavy Metal Stress

Though the rate of metal deposition like Cd, Pb, Hg, As, Zn, etc. at the global level has declined over the past decades, the bioconcentration effect still exerts toxic effects on soil microorganisms due to localized or globalized changes (Wu et al. 2010). Under in vitro conditions detrimental effects of Cd, Pb, and Zn on critical life stages of AMF including spore germination, presymbiotic hyphal extension and sporulation, and symbiotic sporulation were observed (Pereira et al. 2014). However it was also seen that under the same low metal conditions, *Glomus etunicatum* and *Glomus intraradices* showed different levels of sensitivity to metals and later were less sensitive to all three metals (Aguilera et al. 2014). Presymbiotic hyphal extension increases in *G. intraradices* in Cd and Pb solutions (Zhuang et al. 2014). Further the toxicity of higher metal concentrations on symbiotic mycelial expansion and symbiotic sporulation were limited to the metal-enriched compartment in *G. intraradices* (Kruger et al. 2012). It was also observed that toxic effects were more evident on symbiotic sporulation than symbiotic mycelium expansion (Khade and Adholeya 2009).

4.7 Ecological Development of the Rhizosphere by AMF

The rhizosphere is referred as the zone of higher microbial abundance represented by few millimeters of soil inside/outside the roots of plants influenced mainly by root exudates (Tan et al. 2011). It is characterized by sharp gradients of microbial abundance and chemical characteristics (Helgason and Fitter 2009). The highly dynamic nature of rhizosphere or plant-driven microenvironment highly considered by biogeochemical processes, i.e., interactions between plant root processes, soil features, and the associated microbes (Utobo et al. 2011). The annual flux of carbon and other nutrients is mostly maintained by soil communities at the top 5–10 cm of the soil profile and is therefore essential for ecosystems functioning (Ferreira et al. 2015). Though it is widely accepted about the role of soil microbes for buffering against various environment stresses, but still no complete inventory of soil microbes is available (Xun et al. 2015). As mentioned rhizosphere microorganisms particularly AMF allow the growth and survival of plant species on contaminated areas like mine sites by toxic metal immobilization (Liu et al. 2015). These fungi contribute directly to plant growth and survival in contaminated soils by hyphae binding of metals in the rhizosphere (Redecker et al. 2003). And these AMF-associated plants showed a higher degree of metal resistance or tolerance (Verbruggen et al. 2012). Various groups of fungi from major taxonomic groups are reported to thrive in HM-contaminated areas and are able to help other plants to grow for phytoremediation purposes (Holland et al. 2014). The overall mechanisms like reducing abundance and species diversity or selection species-specific resistant population are numerous and varied in their action (Gadd 2010; Jiao et al. 2011). Various reports mention that mycorrhizal associations improve the toxicity of heavy metals to higher plants (Wu et al. 2010; Hossain et al. 2012). *Scleroderma citrinum*, *Paxillus involutus*, and *Laccaria laccata* have no effect on *Pinus sylvestris*. But *Suillus bovinus* and *S. luteus* increases the Cd tolerance (Gamalero et al. 2003). Similarly, Hassan et al. (2015) showed that *Scleroderma flavidum* increase Ni tolerance of *Betula papyrifera*, and none of the other fungal species increased Cu protection. In another report, Sanjay and Kumar (2011) found that Zn toxicity to *Betula pendula* was improved AMF association, which was positively correlated with the degree of compatibility between the fungal strain (*Paxillus involutus*) and the plant, and it was observed that the colonization of *Festuca rubra* and *Calamagrostis epigejos* by AMF decreased Zn toxicity. Further it is believed that herds of cows are impounded with electric fence on mine sites or metal-contaminated soils by providing fodder and water (Kikoti 2013). The cows not only stabilize the upper soil by their hoofs but also enrich soil with urination besides added microbes to the soil through the dung; thus, cattle have a definite role in rhizosphere development vis-à-vis improve plant root associations (Pineiro et al. 2010).

References

Aguilera P, Cornejo P, Borie F, Barea JM, von Baer E, Oehl F (2014) Diversity of arbuscular mycorrhizal fungi associated with *Triticum aestivum* L. plants growing in an Andosol with high aluminum level. Agric Ecosyst Environ 186:178–184

Amal AMA, Hasnah MJ (2012) Interaction between arbuscular mycorrhiza and heavy metals in the rhizosphere and roots of Juniperus procera. Int J Agric Biol 14(1):69–74

Ashraf MA, Maah MJ, Yusoff I (2013) Evaluation of natural phytoremediation process occurring at extin mining catchment. Chiang Mai Journal of Science 40:198–213

Azevedo RA, Gratão PL, Monteiro CC, Carvalho RF (2012) What is new in the research on cadmium-induced stress in plants? Food Energy Secur 1:133–140

Banerjee S, Chandel G (2011) Understanding the role of metal homeostasis related candidate genes in Fe/Zn uptake, transport and redistribution in rice using semi-quantitative RT-PCR. Journal of Plant Molecular Biology & Biotechnology 2(1):33–46

Bonanno G, Giudice LR (2010) Heavy metal bioaccumulation by the organs of Phragmites australis (common reed) and their potential use as contamination indicators. Ecol Indic 10(3):639–645

Bussche JV, Soares EV (2011) Lead induces oxidative stress and phenotypic markers of apoptosis in Saccharomyces cerevisiae. Appl Microbiol Biotechnol 90:679–687

Cao LX, Jiang M, Zeng ZR, Du AX, Tan HM, Liu YH (2008) *Trichoderma atroviride* F6 improves phytoextraction efficiency of mustard (*Brassica juncea* (L.) Coss. var. *foliosa* Bailey) in Cd, Ni contaminated soils. Chemosphere 71:1769–1773

Chaudhry MS, Batool Z, Khan AG (2005) Preliminary assessment of plant community structure and arbuscular mycorrhizas in rangeland habitats of Cholistan Desert, Pakistan. Mycorrhiza 15(8):606–611

Chen Z, Tang J, Zhi G, Hu S (2005) Arbuscular mycorrhizal colonization and phosphorus acquisition of plants: effects of coexisting plant species. Appl Soil Ecol 28:259–269

Dorota AS, Markiewicz J, Wolf WM (2015) Heavy metal uptake by herbs. IV. Influence of soil pH on the content of heavy metals in *Valeriana officinalis* L. Water Air Soil Pollut 226(4):106

Ebrahim S, Mojtaba K, Mohammadreza S, Ghasem HT, Hassan S (2014) Microorganism's application strategy for bio-phytoremediation of heavy metal: a review. J Bio & Env Sci 5(3):289–298

Elekes CC (2014) Eco-technological solutions for the remediation of polluted soil and heavy metal recovery. In: Hernández-Soriano MC (ed) Environmental risk assessment of soil contamination. InTech, Rijeka, pp 309–335

Ferreira PAA, Ceretta CA, Soriani HH, Tiecher TL, Soares CRFS, Rossato LV (2015) *Rhizophagus clarus* and phosphate alter the physiological responses of *Crotalaria juncea* cultivated in soil with a high Cu level. Appl Soil Ecol 91:37–47

Fijalkowski K, Kacprzak M, Grobelak A, Placek A (2012) The influence of selected soil parameters on the mobility of heavy metals in soils. Inzynieria i Ochrona Srodowiska 5:81–92

Gadd GM (2010) Metals, minerals and microbes: geomicrobiology and bioremediation. Microbiology 156:609–643

Gamalero E, Trotta A, Massa N, Copetta A, Martinotti MG, Breta G (2003) Impact of two fluorescent pseudomonads and an arbuscular mycorrhizal fungus on tomato plant growth, root architecture and P acquisition. Mycorrhiza 14(3):185–192

Gao Y, Cheng Z, Ling W, Huang J (2010) Arbuscular mycorrhizal fungal hyphae contribute to the uptake of polycyclic aromatic hydrocarbons by plant roots. Bioresour Technol 101:6895–6901

Gonzalez-Chavez MC, Carrillo-Gonzalez R, Wright SF, Nichols K (2004) The role of glomalin, a protein produced by arbuscular mycorrhizal fungi, in sequestering potentially toxic elements. Environ Pollution 30(3):317–323

Gosling P, Mead A, Proctor M, Hammond JP, Bending GB (2010) Contrasting arbuscular mycorrhizal communities colonizing different host plants show a similar response to a soil phosphorus concentration gradient. New Phytol 198:546–556

Guan JC, Yeh CH, Lin YP, Ke YT, Chen MT, You JW, Liu YH, Lu CA, Wu SJ, Lin CY (2010) A 9 bp cis-element in the promoters of class I small heat shock protein genes on chromosome 3 in rice mediates L-azetidine-2-carboxylic acid and heat shock responses. J Exp Bot 61:4249–4261

Guerinot ML (2010) Iron in cell biology of metals and nutrients. In: Hell R, Mendel R (eds) Plant cell monographs, vol 17. Springer. Berlin/Heidelberg, pp 75–94

Hall JL (2002) Cellular mechanisms for heavy metal detoxification and tolerance. J Exp Bot 53:1–11

Hassan Z, Aarts MGM (2011) Opportunities and feasibilities for biotechnological improvement of Zn, Cd or Ni tolerance and accumulation in plants. Environ Exp Bot 72:53–63

Hassan MA, Rouf AR, Tiralongo E, Tom WM, Tiralongo J (2015) Mushroom lectins: specificity, structure and bioactivity relevant to human disease. Int J Mol Sci 16:7802–7838

Hassinen VH, Tervahauta AI, Schat H, Kärenlampi SO (2011) Plant metallothioneins–metal chelators with ROS scavenging activity? Plant Biol 13:225–232

Helgason T, Fitter AH (2009) Natural selection and the evolutionary ecology of the arbuscular mycorrhizal fungi (Phylum Glomeromycota). J Exp Bot 60:2465–2480

Hildebrandt U, Regvar M, Bothe H (2007) Arbscular mycorrhiza and heavy metal tolerance. Phytochemistry 68:139–146

Holland TC, Bowen P, Bogdanoff C, Hart MM (2014) How distinct are arbuscular mycorrhizal fungal communities associating with grapevines? Biol Fertil Soils 50:667–674

Hossain MA, Piyatida P, da Silva JAT, Fujita M (2012) Molecular mechanism of heavy metal toxicity and tolerance in plants: central role of glutathione in detoxification of reactive oxygen species and methylglyoxal and in heavy metal chelation. Journal of Botany 2:1–37: Article ID 872875

Hussain D, Haydon M, Wang Y, Wong E, Sherson S, Young J, Camakaris J, Harper J, Cobbett C (2004) P-type ATPase heavy metal transporters with roles in es- sential zinc homeostasis in Arabidopsis. Plant Cell 16:1327–1339

Irena S, Ajit V (2011) Detoxification of heavy metals. Springer, Heidelberg/Dordrecht/London/New York, p 19

Jia Y, Huang H, Sun G, Zhao FJ (2012) Pathways and relative contributions to arsenic volatilization from rice plants and paddy soil. Environ Sci Technol 46:8090–8096

Jiang CA, Wu QT, Sterckeman T, Schwartz C, Sirguey C, Ouvrard S, Perriguey J, Morel JL (2010) Co-planting can phytoextract similar amounts of cadmium and zinc to mono-cropping from contaminated soils. Ecol Engineer 36:391–395

Jiao H, Chen YL, Lin XG, Liu RJ (2011) Diversity of arbuscular mycorrhiza fungi in greenhouse soils continuously planted to watermelon in North China. Mycorrhiza 21:681–688

Joner EJ, Briones R, Leyval C (2000) Metal-binding capacity of arbuscular-mycorrhizal mycelium. Plant Soil 226(2):227–234

Joschim HJ, Makoi R, Ndakidemi PA (2009) The agronomic potential of vesicular-arbuscular mycorrhiza (AM) in cereals– legume mixtures in Africa. Afr J Microbiol Res 3(11):664–675

Khade SW, Adholeya A (2009) Arbuscular mycorrhizal association in plants growing on metal-contaminated and noncontaminated soils adjoining kanpur tanneries, Uttar Pradesh, India. Water Air Soil Pollut 202:45–56

Kikoti IA (2013) The impacts of livestock grazing on vegetation in Montane forest of the northern slopes of Mount Kilimanjaro. Unpublished master's thesis, University of Dares salaam, DaresSalaam

Kramer U, Talke IN, Hanikenne M (2007) Transition metal transport. FEBS Lett 581:2263–2272

Kruger M, Kruger C, Walker C, Stockinger H, Schußler A (2012) Phylogenetic reference data for systematics and phylotaxonomy of arbuscular mycorrhizal fungi from phylum to species level. New Phytol 193:970–984

Lavania D, Dhingra A, Siddiqui MH, Al-Whaibi MH, Grover A (2015) Current status of the production of high temperature tolerant transgenic crops for cultivation in warmer climates. Plant Physiol Biochem 86:100–108

Lee LY, Fang MJ, Kuang LY, Gelvin SB (2008) Vectors for multicolor bimolecular fluorescence complementation to investigate protein-protein interactions in living plant cells. Plant Methods. doi:10.1186/1746-4811-4-24

Leung HM, Ye ZH, Wong MH (2007) Survival strategies of plants associated with arbuscular mycorrhizal fungi on toxic mine tailings. Chemosphere 66:905–915

Lin X, Feng Y, Zhang H, Chen R, Wang J, Zhang J (2012) Long-term balanced fertilization decreases arbuscular mycorrhizal fungal diversity in an arable soil in North China revealed by 454 pyrosequencing. Environ Sci Technol 46:5764–5771

Liu Z, Li H, Soleimani M (2008) Cd^{2+} versus Zn^{2+} uptake by the ZIP8 HCO_3—dependent symporter: kinetics, electrogenicity and trafficking. Biochem Biophys Res Commun 365(4):814–820

Liu H, Yuan M, Tan S, Yang X, Lan Z, Jiang Q (2015) Enhancement of arbuscular mycorrhizal fungus (*Glomus versiforme*) on the growth and Cd uptake by Cd-hyperaccumulator *Solanum nigrum*. Appl Soil Ecol 89:44–49

Long LK, Yao Q, Guo J, Yang RH, Huang YH, Zhu HH (2010) Molecular community analysis of arbuscular mycorrhizal fungi associated with five selected plant species from heavy metal polluted soils. Eur J Soil Biol 46:288–294

Ma X, Richter AR, Albers S, Burken JG (2005) Phytoremediation of MTBE with hybrid poplar trees. Int Phytoremediation 6(2):157–167

Malik RN, Husain SZ, Nazir I (2010) Heavy metal contamination and accumulation in soil and wild plant species from industrial area of Islamabad, Pakistan. Pak J Bot 42:291–301

Mganga N, Manoko MLK, Rulangaranga ZK (2011) Classification of plants according to their heavy metal content around north Mara gold mine, Tanzania: implication for phytoremediation. Tanzania Journal of Science 37:109–119

Miransari M (2011) Hyperaccumulators, arbuscular mycorrhizal fungi and stress of heavy metals. Biotechnol Adv 29(6):645–653

Mourato M, Reis R, Martins L (2012) Characterization of plant antioxidative system in response to abiotic stresses: a focus on heavy metal toxicity. In: Montanaro G, Dichio B (eds) Advances in selected plant physiology aspects. Intech, Rijeka, pp 23–44

Neumann D, Lichtenberger O, Günther D, Tschiersch K, Nover L (1994) Heat-shock proteins induce heavy-metal tolerance in higher plants. Planta 194:360–367

Oh K, Cao T, Li T, Cheng H (2014) Study on application of phytoremediation technology in management and remediation of contaminated soils. Journal of Clean Energy Technologies 2:216–220

Park JD (2010) Heavy metal poisoning. Hanyang Med Rev 30:319–325

Pereira CMR, da Silva DKA, de Almeida Ferreira AC, Goto BT, Maia LC (2014) Diversity of arbuscular mycorrhizal fungi in Atlantic forest areas under different land uses. Agric Ecosyst Environ 185:245–252

Pineiro G, Paruelo JM, Oesterheld M, Jobbagy EG (2010) Pathways of grazing effects on soil organic carbon and nitrogen. Rangel Ecol Manag 63:109–119

Redecker D, Hijri I, Wiemken A (2003) Molecular identification of arbuscular mycorrhizal fungi in roots: perspectives and problems. Folia Geobotanica 38:113–124

Rees F, Simonnot MO, Morel JL (2014) Short-term effects of biochar on soil heavy metal mobility are controlled by intra-particle diffusion and soil pH increase. Eur J Soil Sci 65(1):149–161

Richau KH, Schat H (2009) Intraspecific variation of nickeland zinc accumulation and tolerance in the hyperaccumulatoro *Thlaspi caerulescens*. Plant Soil 314:253–262

Salt DE, Kramer V (2000) Mechanism of metal hyperaccumulation in plants. In: Raskin I, Ensley BD (eds) Phytoremediation of toxic metals using plants to clean up the environment. Wiley, New York, pp 231–246

Salt DE, Smith RD, Raskin I (1998) Phytoremediation. Annu Rev Plant Physiol Plant Mol Biol 49:643–668

Sanjay KJHA, Kumar N (2011) Potential of mycorrhizal fungi in ecosystem: a review. International Journal of Research in Botany 1(1):1–7

Sarkar NK, Kundnani P, Grover A (2013) Functional analysis of Hsp70 superfamily proteins of rice (Oryza sativa). Cell Stress Chaperon 18(4):427–437

Satish AB (2013) Arbuscular mycorrhizal fungi: a potential biotechnological tool for phytoremediation of heavy metal contaminated soils. IJSN 4(1):1–15

Schutzendubel A, Polle A (2002) Plant responses to abiotic stresses: heavy metal-induced oxidative stress and protection by mycorrhization. The Journal of Experimental Botany 53(372):1351–1365

Shakeel S, Haq NU, Heckathorn SA, Hamilton EW, Luthe DS (2011) Ecotypic variation in chloroplast small heat-shock proteins and related thermotolerance in Chenopodium album. Plant Physiol Biochem 49:898–908

Shirmohammadi E, Khaje M, Shirdali M, Hosein GT, Shahgholi H (2014) Microorganism's application strategy for bio-phytoremediation of heavy metal: a review. Journal of Biodiversity and Environmental Sciences (JBES) 5(3):289–298

Siddiqui MM, Abbasi BH, Ahmad N, Ali M, Mahmood T (2014) Toxic effects of heavy metals (Cd, Cr and Pb) on seed germination and growth and DPPH-scavenging activity in Brassica rapa var. Turnip. Toxicol Ind Health 30:238–249

Singh PK (2012) Role of glomalin related soil protein produced by arbuscular mycorrhizal fungi: a review. Agric Sci Res J 2(3):119–125

Singh J, Kalamdhad AS (2013) Chemical speciation of heavy metals in compost and compost amended soil—a review. International Journal of Environmental Engineering Research 2:27–37

Tan F, Wang J, Chen Z, Feng Y, Chi G, Rehman SU (2011) Assessment of the arbuscular mycorrhizal fungal community in roots and rhizosphere soils of Bt corn and their non-Bt isolines. Soil Biol Biochem 43:2473–2479

Utobo EB, Ogbodo EN, Nwogbaga AC (2011) Techniques for extraction and quantification of arbuscular mycorrhizal fungi. Libyan Agriculture Research Center Journal International 2:68–78

Vamerali T, Bandiera M, Mosca G (2010) Field crops for phytoremediation of metal-contaminated land. A review. Environ Chem Lett 8:1–17

Verbruggen N, Hermans C, Schat H (2009) Molecular mechanisms of metal hyperaccumulation in plants. New Phytol 182:781–781

Verbruggen E, Van Der Heijden MG, Weedon JT, Kowalchuk GA, Roling WF (2012) Community assembly, species richness and nestedness of arbuscular mycorrhizal fungi in agricultural soils. Mol Ecol 21:2341–2353

Wang X, Winter D, Ashrafi G, Schlehe J, Wong YL, Selkoe D, Rice S, Steen J, LaVoie MJ, Schwarz TL (2011) PINK1 and Parkin target Miro for phosphorylation and degradation to arrest mitochondrial motility. Cell 147(4):893–906

Williams LE, Pittman JK, Hall JL (2000) Emerging mechanisms for heavy metal transport in plants. Biochim Biophys Acta 1465(1–2):104–126

Wu FY, Bi YL, Leung HM, Ye ZH, Lin XG, Wong MH (2010) Accumulation of As, Pb, Zn, Cd and Cu and arbuscular mycorrhizal status in populations ofCynodon dactylon grown on metal-contaminated soils. Appl Soil Ecol 44:213–218

Wuana RA, Okieimen FE (2011) Heavy metals in contaminated soil: a review of sources, chemistry, risks and best available strategies for bioremediation. ISRN Ecol 1–20

Xun F, Xie B, Liu S, Guo C (2015) Effect of plant growth-promoting bacteria (PGPR) and arbuscular mycorrhizal fungi (AMF) inoculation on oats in saline-alkali soil contaminated by petroleum to enhance phytoremediation. Environ Sci Pollut Res 22:598–608

Yang Y, Song Y, Scheller HV, Ghosh A, Ban Y, Chen H (2015) Community structure of arbuscular mycorrhizal fungi associated with *Robinia pseudoacacia* in uncontaminated and heavy metal contaminated soils. Soil Biol Biochem 86:146–158

Zarei M, Hempel S, Wubet T, Schafer T, Savaghebi G, Jouzani GS, Nekouei MK, Buscot F (2010) Molecular diversity of arbuscular mycorrhizal fungi in relation to soil chemical properties and heavy metal contamination. Environ Pollut 158:2757–2765

Zeng F, Ali S, Zhang H, Ouyang Y, Qiu B, Wu F, Zhang G (2010) The influence of pH and organic matter content in paddy soil on heavy metal availability and their uptake by rice plants. Environ Pollut 159:84–91

Zhuang P, Lu H, Li Z, Zou B, McBride MB (2014) Multiple exposure and effects assessment of heavy metals in the population near mining area in South China. PLoS One 9(4), e94484

Phytoremediation: A Green Technology

5

Contents

Abstract

Phytoremediation is a lucrative plant-based advancement to remediation that takes lead in its knack to degrade environmental contaminants and to metabolize diverse pollutants. Heavy metals and organic pollutants are the central candidates for phytoremediation. To eliminate contaminants from soil, sediment, and/or water, plants can degrade organic pollutants or restrain and alleviate metal pollutants by substituting as filters or traps. Plants are exclusive organisms fitted with significant metabolic and absorption competence, plus transport systems that translocate nutrients or pollutants exclusively from the growth milieu, soil, or water. An in-depth perceptive of the physiological and molecular process of phytoremediation emerged as an efficient biological and engineering stratagem intended to optimize and enhance the prospective of phytoremediation process. Besides, various field

S. Jan, J.A. Parray, *Approaches to Heavy Metal Tolerance in Plants*,
DOI 10.1007/978-981-10-1693-6_5

experiments established the viability of using plants for cleaner environment. Metal-contaminated waters and soils are a main ecological crisis, and the majority of conventional remediation methods do not present adequate exposition. Utilization of genetically tailored plants is particularly selected for more efficient phytoremediation. In this chapter, we will illustrate numerous biological processes of heavy metal uptake, translocation, and resistance plus approaches for enhancing phytoremediation potential. Commercialization of bioremediation via genetically engineered microbes is contorted by debate over the expertise; transgenic plants may take center stage for environmental cleanup.

Keywords
Contaminants • Bioremediation • Phytostabilization • Phytoextraction • Phytovolatilization

5.1 Introduction

Phytoremediation is defined as the utilization of plants for in situ management of polluted soils, sediments, and water. It is an ecologically safe biotechnological intervention that has grabbed attention of scientists, civic view, regulators, and municipal organization (Marmiroli et al. 2006). Phytoremediation is broad term consisting of the Greek prefix *phyto* (plant), attached to the Latin root *remedium* (to eliminate an iniquity) (Tangahu et al. 2011). Commonly, phytoremediation is defined as the promising prospective via specific plants to clean up the polluted environment from dangerous pollutant to advance the environment quality. The methods for cleaning organic contaminants entail phytostabilization, rhizodegradation, rhizofiltration, phytodegradation, and phytovolatilization (Sheoran et al. 2012). The root plants exudate to alleviate, discharge, and combine the contaminants in the soil matrix, thus dropping their bioavailability (Wuana and Okieimen 2011). These are together termed as phytostabilization process. Some crops have been exploited to accumulate pollutants from soil and groundwater via absorption and adsorption onto roots or rainwater inside the root zone (Prasad and Freitas 2003). This process is for organic and metal pollutants in soils, sediments, and sludge medium. Specific crops hyperaccumulate metal pollutants and/or supplemented nutrients in harvestable root and shoot tissue, from the growth substrate through phytoextraction mechanism.

Phytovolatilization is capacity to take up and subsequently mobilize the pollutants into the atmosphere (Kvesitadze et al. 2006). While phytotransformation/phytodegradation mechanism is the degradation of pollutants absorbed by plants via metabolic system inside the plant or the degradation of pollutants superficially to the plant by the influence of components generated via plants (Parmar and Singh 2015). This system is for intricate organic pollutants that are broken into smaller organic contaminants in soils, sediments, sludges, and groundwater medium (Ali et al. 2013). Plant roots absorb metal pollutants and/or excess compounds from growth substrates through rhizofiltration mechanism, adsorption, or rainfall onto plant roots or assimilation into the roots of pollutants that are present in soil sap around root zone (Yadav et al. 2011). This mechanism is meant for metals, excess minerals, and radioactive contaminants in diverse water resources. The degradation of pollutants in the soil sap via microbial process that is enhanced through subsistence of the root zone is known as rhizodegradation (Lu et al. 2011). This process utilizes microbes to utilize and absorb organic constituents for sustenance and energy. Organic matter released via plant roots, sugars, alcohols, and acids possesses organic carbon that contribute food for soil microbes and establish compact root mass that absorbs enormous water (Maier et al. 2009). This mechanism is for organic pollutants in soil.

5.2 Phytoextraction

Phytoextraction is absorption and transportation of pollutants through plant rhizosphere over soil segment of the plant shoot that must be produced and burned attaining energy and recovering the metal from the ash (Zaier et al. 2010). Phytoextraction or phytoaccumulation is metal uptake from soil via plant rhizosphere into shoot of plants. Hyperaccumulators are those plants which absorbs toxic metals above their normal levels and after attaining permissble growth, they are harvested and either incinerates or composted to recycle the metals (Prasad 2013). This practice may be continual as indispensable to carry soil contaminant levels down to acceptable limits. If plants are burned, the resulting ash can be dumped off in waste landfill, but the capacity of ash should be lower than 10 % of the quantity that would be produced if the polluted soil itself were dug up for organization (Rajkumar et al. 2010). Metals, for instance, nickel, zinc, and copper, are the superlative entrant for exclusion through phytoextraction since most of the almost 400 known crops that take up extraordinarily huge quantity of metals have an elevated similarity for accruing these metals. Plants that assimilate lead and chromium are presently being investigated and analyzed (Babaeian et al. 2016; Escobar and Dussán 2016).

5.3 Phytostabilization

Phytostabilization is the utilization of selected crops to block the pollutants in the soil and groundwater through assimilation and accretion in crops, adsorption onto roots, or rainfall around the root zone avoiding their translocation in soil, together with their transport via erosion and devaluation (Mertens et al. 2004). These processes decrease the movement of the pollutant and avoid passage to the groundwater or air and also decline mobility for entry into the food chain. Hyperaccumulators are utilized for restoration of vegetation sites, hence decreasing the expected passage of pollution through wind erosion and reclamation of surface soils and leaching of soil pollution to groundwater (Tordoff et al. 2000).

5.4 Rhizofiltration

Rhizofiltration ("rhizo" means "root") is the process of adsorption and sequestration of pollutants around the root zone in soil sap to clean public wastewater (Yadav et al. 2011; Rawat et al. 2012). Rhizofiltration is analogous to phytoextraction, although crops are utilized for remediation of contaminated groundwater instead of soil. The crops are utilized for remediations which are raised under greenhouse. Wani et al. (2012) demonstrated remediation of polluted sites through the collection of wastewater from polluted sites through roots. Plant roots after accumulating pollutants were harvested. Vanek et al. (2010) reported that sunflowers can degrade radioactive pollutants from pond water in a test at Chernobyl, Ukraine.

5.5 Phytovolatilization

Phytovolatilization is the absorption and volatilization of pollutants by a plant, with release of the contaminant or a modified form of pollutant into the atmosphere. Phytovolatilization has been practiced in growing trees which adsorb water together with contaminants (Sakakibara et al. 2010). Few of the pollutants can translocate to leaves and mobilize into the atmosphere at moderately lesser concentrations. Plants perform secondary role by adhering pollutants in their rhizosphere from soil sap, averting soil erosion and conserving the soil surface, and reducing the effects of rain. Concurrently, plant roots release nutrients that regulate luxuriant microbial diversity in the rhizosphere. Microbial diversity in rhizosphere is influenced through interactions among soil type, plant species, and root zone sites. Microbial diversity is usually elevated in the root zone than in the root-free soil. This is owing to symbiotic association among soil microbes and plants. This symbiotic relationship can enhance diminutive bioremediation mechanism. Plant roots can also offer root surface for absorption or rainfall of metal pollutants (Bolan et al. 2014). For phytoremediation, the rhizosphere is of an outstanding implication. The pollutants can be adsorbed through root which subsequently accrue and volatilize through plant surface. Metabolization of pollutants in the soil by plant enzymes that originate from the roots is an additional phytoremediation process (Jabeen et al. 2009). For several pollutants, passive uptake through micropores in the root cell walls might be chief route into the root, where degradation occurs (Ansari et al. 2013).

5.6 Fundamental Mechanism of Heavy Metals and Inorganic Contaminant Uptake and Transport

Rhizosphere exists as main platform for dialogue between plants and metals. Most of the phytotechnologies are applicable only when pollutant exists in contact with rhizosphere and most of them depend on heavy metal uptake via roots. This is the main reason for subsistence of plasma membrane transporters as main premise to explore for the proper implementation of phytotechnology. Heavy metal uptake involved the same kind of transporters which are utilized for macro- and micronutrient transport. Perfus-Barboech et al. (2002) have confirmed the involvement of Ca channels in Cd absorption in *Arabidopsis thaliana*. The viewpoint of Cd that mimics Ca in plant cells can also authenticate its toxicity with deterioration of metabolism and homeostasis of this essential element. Zhang et al. (2013) evaluated plant cell protoplasts to confirm if discrepancy in transport in sink tissues could elucidate the varied performance of hyperaccumulators. The results obtained with *A. halleri* and *Thlaspi caerulescens* exhibit that leaf plasma membranes do not exhibit discrepancy in transport (Cosio et al. 2004). Subsequently, it has been speculated that different processes might be instant to escort the metals to their subcellular sections as vacuoles where in they exist as reserve, e.g., lignocellulose. Soudek et al. (2004) illustrated transport of radionuclides which can advance from autoradiographic techniques with Cs. Imaging techniques potentiate contrast among diverse species for metal uptake efficiency, but they have also demonstrated possible sink tissues supplying imperative data for execution of phytoextraction.

5.6.1 Accumulation and Sequestration

In order to elucidate mechanism of hyperaccumulation, several researchers have illustrated difference between hyperaccumulator and non-accumulator crop species to identify localization and speciation of heavy metals within plant species (Donner et al. 2012). Sarret et al. (2002) reported sequestration of Zn in vacuoles of leaf trichomes and mesophyll cells of *Arabidopsis halleri*, known as Zn and Cd hyperaccumulator. This research has illustrated that roots delineated Zn through malate and phytate or possibly phosphate, whereas in trichomes it is coordinated by C atoms related to organic acids as analyzed by EXAFS. However, it has been confirmed that Zn cannot accumulate in trichomes so it cannot be marked as main sink. Moreover, Zn binding with phosphate has also been implicated in non-accumulator *Arabidopsis lyrata*. To understand molecular source of the hyperaccumulation capability and to describe the accretion strategies will be involved in advancing and execution of phytoextraction. Marmiroli et al. (2004) demonstrated Ni accretion in the leaf trichomes of *Alyssum bertolonii* as compared to non-accumulator *Alyssum montanum* that stores Ni in roots as analyzed via scanning electron microscopy and microanalysis. Marmiroli et al. (2005) illustrated accretion of Pb in roots of walnut trees where Pb forms complex with carbon skeleton of cellulose and lignin as analyzed by EXAFS. Rubin et al. (2000) has employed multiple approaches to explore the mechanism of hyperaccumulation through varied disciplines like genomics and proteomic strategies. In order to investigate and discover new sequences and identify novel molecular markers and orthologous genes, we need functional genomics and proteomics to reach some objective. Ectopic expression of *Thlaspi* genes in yeast consequently resulted in segregation of gene P-type ATPase that is implicated in Cd transport (Bernard et al. 2004).

Genetic mapping is the process of inclination regarding quantitative traits, and analysis is being carried out for building maps of quantitative trait loci (QTLs) for hyperaccumulation. *Brassicaceae* is the most illustrated family having hyperaccumulators like *Arabidopsis thaliana* with completely genomic sequence available on genetic database. Payne et al. (2004) has demonstrated extensive QTL map for *Arabidopsis thaliana* exhibiting Cs accretion. Several genomic accessions have been scrutinized for Cs accretion advancing to explication of double-fold discrepancy in Cs accumulation. Crosses between divergent phenotypes and investigation in segregating progenies progress toward the mapping of recognized QTLs on few chromosomes; the subsistence of two QTLs on chromosomes I and V was confirmed via investigation of segregating populations from autonomous crosses (Takagi et al. 2013). Mapping of candidate genes in specific regions would advance toward new premise regarding the configuration and utility of these QTLs (Buescher et al. 2010).

5.6.2 Hereditary Basis of Tolerance

Classic genetic analysis has been dominated successively to cope with the issue of genetic source of tolerance and/or accumulation in addition to genomic and proteomic progress toward gene identification. In this pretext, *Arabidopsis halleri* and *Thlaspi caerulescens* exist as model plants for being two best hyperaccumulators which can be crossed with non-hyperaccumulator ecotypes for assessment of characteristic segregation (Rascio and Navari-Izzo 2011). Genetic evaluation has confirmed that tolerance to heavy metal is highly specific and is polygenic trait. Craciun et al. (2006) illustrated interspecific crosses between *A. halleri* and *A. lyrata* ssp. *petraea* which supplemented information regarding tolerance and hyperaccumulation to Cd. Tolerance and hyperaccumulation are segregated as autonomous traits, whereas Cd and Zn tolerance segregate codependently (Sasaki et al. 2014). Furthermore, Cd and Zn hyperaccumulation are co-regulated or executed through similar genes (Antosiewicz et al. 2013). The similar strategy has been trailed in *Thlaspi* with ecotypes incompatible in accretion ability. Segregation results suggest that Cd accumulation is polygenic trait and Zn accumulation is governed by two genes. Relationship among accumulation of Zn, Cd and Mn is invariable with diverse transporter having instant specificity for the three metals. Cd tolerance and accumulation segregate as independent characteristics in *Arabidopsis thaliana*.

5.7 Basic Mechanisms: Organic Contaminants

5.7.1 Mechanisms of Genetic Controls: Candidate Genes

Modern research has been targeted toward specific gene identification using genomic approach. Presently, researchers are investigating explicit candidate genes encoding enzymes responsible for degradation of pollutants and involved in metabolism of xenobiotics. The exceptionally recognized paradigm is glutathione transferase (GST), a multigenic family lately investigated and analyzed in rice. Plant genomes possess 25–60 GST genes. Five diverse types are identified in plant taxa, two being plant specific, namely, Phi and Tau (Sappl et al. 2004). The chief significance of these designed for phytotechnologies is the understanding of glutathione–xenobiotic conjugates in secondary stage II of the pollutant metabolism. Leoni et al. (2011) genetically analyzed separation and identification of GST genes in significant crops as *Arabidopsis*, rice, maize, and soybean. Rice possesses 61 genes out of which 55 % expressed via EST libraries with tau class being most expressed. However, several GST enzymes are stimulated and expressed under exact set of factors. In the case of rice about 30 GST genes that are grouped together on chromosome 10, similar feature has been exhibited in other crops (Jain et al. 2010). Soranzo et al. (2004) confirmed the evolutionary development of Phi and Tau in response to stress induced by toxic pollutants. Some crop species exhibited coevolution of diverse enzymes and GST involved in xenobiotic metabolism particularly cytochromes P450 and ABC transporters can be enumerated as significant defense entrants against varied pollutants (Dauchy et al. 2008).

5.7.2 Investigation and Classification of Enzymes and Proteins

Progress in proteomic research has led to identification of new enzymes involved in xenobiotic metabolism. Su et al. (2012) demonstrated segregation of glucosyltransferase from *Arabidopsis thaliana* responsible for the degradation of 3,4-dichloroaniline. After purification from in vitro cell cultures, the enzyme was analyzed with MALDI-TOF MS, and cloning of the gene was likely dependent on sequence data. Stimulation of the enzyme by herbicide safeners provides elucidation of interactions between xenobiotics and plant metabolism.

5.7.3 Transgenic Strategies

Modulation in metabolism and detoxification of xenobiotic compounds is dependent on single enzyme. Sinha et al. (2010) illustrated advantages of transgenic strategy for modulating or improving such enzymes to implement phytotechnology. Singh et al. (2011) reported potential of transgenic engineering in advancing herbicide tolerance by modulating gene expression of cytochrome P450 involved in phase I of the metabolism. Didierjean et al. (2002) transferred gene from Jerusalem artichoke stimulated by metals and drugs; it imparts tolerance to phenylurea in sensitive crops as tobacco and *Arabidopsis*. Chemical pollution of soil and water resources by herbicides can be subsided by this gene through phytotransformation. The succeeding illustration is not related only to phytoremediation but to phytomonitoring of organic pollutants using transgenic strategy. Danish corporation Aresa Biotechnology has advanced GM plants like *Arabidopsis thaliana* that recognize nitrogen dioxide release via explosives and transmit this signal through change to red color (Daniell 2002). The suggested significance includes plant cultivation in contaminated areas to ensure remediation of particular location. Cell cultures are utilized for evaluation before its useful implication in constructed wetlands, mainly in the case of complex pollutants. Marmiroli et al. (2006) reported disparity in species pertaining to metal uptake and detoxification potential like *Carex* has irrelevant TNT uptake as compared to *Juncus*, *Phragmites*, or *Typha*.

5.7.4 Metal Transporters and Interactions in Membranes at Molecular Level

Micronutrients such as iron and zinc are required for normal growth in both plants and animals, while accretion of nonessential metals like cadmium or lead can be extremely detrimental. Proteins of cation diffusion facilitator now articulated as citation efflux family are involved in the homeostasis of Cd^{2+}, Co^{2+}, Fe^{2+}, and Zn^{2+} in microbes, animals, and plants (Williams et al. 2000). Consequently, explanation of the role of CDF proteins in *Arabidopsis thaliana* would be advantageous to the attainment of phytoremediation. Corresponding DNAs are to be functionally expressed in appropriate mutants of *Saccharomyces cerevisiae* to verify their function. Persans et al. (2001) reported utilization of representative *Arabidopsis* CDFs in a reverse genetic strategy. Expression of promoters like GUS fusions and epitope-tagged fusion proteins in *A. thaliana* by expansion and exploitation of antibodies regulates and localizes these CDFs (Kirchner 2009). Protein–protein interactions are significant for CDF function dues their dependence on substrate known to be metallochaperone proteins in the cytoplasm.

5.8 Feature Controlling the Metal Uptake

There are several features which can influence the heavy metal uptake mechanism:

5.8.1 Selection of Plant Species

Heavy metal uptake by plants can be evaluated by remediation potential (Weick et al. 2003). The heavy metal uptake is directly governed by plant characteristics. Generally, the phytoextraction process relies on the identification of suitable plant species that hyperaccumulate heavy metals and produce high biomass through known crop productivity and organization practices (Pantola and Alam 2014).

5.8.2 Characteristics of Medium

Agricultural practices are progressed to enhance remediation like pH adjustment, fertilizer, and chelator augmentation (Chen and Cutright 2002). The magnitude of lead uptake by plants is influenced by the pH, organic matter, and phosphorus levels in the soil. Absorption of lead by plants was declined by pH alteration of the soil with lime from level of 6.5 to 7.0 (Emongor 2007).

5.8.3 Rhizosphere

The rhizosphere is of outstanding implication in phytoremediation. Rhizosphere absorbs pollutants and accrues or metabolizes inside plant tissue. Detoxification of contaminants in the soil by plant enzymes originated from the roots is also phytoremediation mechanism. A morphological modification to drought stress is an increase in root diameter and reduced root elongation as a response to lower permeability of the dried soil (Merkl et al. 2005).

5.8.4 Vegetative Uptake

Vegetative uptake is influenced by environmental factors. The temperature influences growth constituents and then root length. Root structure under field setup changes from that of flourishing via greenhouse trial (Bardgett et al. 2014). The completion of phytoremediation especially phytoextraction rely on a pollutant-specific hyperaccumulator (McGrath and Zhao 2003). Considerate mass balance examination and the metabolic upshot of contaminants in plants are the resources to reveal the application of phytoremediation.

Metal uptake by plants relies on the bioavailability of the metal in the water phases, which successively rely on the maintenance stage of the metal added to the boundary line with other elements and substances in the water. Furthermore, when metals are adsorbed on the soil, the pH, redox potential, and organic matter composition will all influence the proclivity of the metal to endure in ionic and plant-available form. Plants alter soil pH and oxygenate the sediment, which affect the convenience of the metals, raising the bioavailability of heavy metals by the accretion of ecological physicochemical factors, such as chelating agents and micronutrients (de Mello-Farias et al. 2011).

5.8.5 Addition of Chelating Agent

The improvement in the heavy metal uptake by the energy crops can be influenced by increasing the availability of heavy metals through accretion of biodegradable physicochemical characteristics, e.g., chelating agents, micronutrients, also by inducing heavy metal uptake ability of the microflora that exist near root zone (Santiago et al. 2004). This instant heavy metal uptake will terminate subsequently less expensive cleaning time span. However, using synthetic chelating agents, the possibility of improved leaching should be considered (Leštan et al. 2008).

The use of chelating agents in heavy metal-polluted soils could support leaching of the pollutants into the soil. Since the availability of heavy metals in soils decline in the range of pH 5.5–6, the exploitation of a chelating agent is defensible and may be essential in alkaline soils. It was confirmed that prolonged exposure of plants to EDTA could improve metal translocation in plant tissue along with general phytoextraction competence. The rationale of a synthetic chelating agent (EDTA) at 5 mmol/kg revealed good records (Alkorta et al. 2004). Plant roots discharge organic acids, e.g., citrate and oxalate, which affect the bioavailability of metals. In chelate mediated phytoremediation, synthetic chelating agents such as NTA and EDTA enhanced phytoextraction potential of soil-polluting heavy metals. The survival of a ligand affects the heavy metal uptake through assemblage of metal–ligand complexes and changes the potential to leach metals around root cap (Nowack et al. 2006).

5.9 Advantages of Phytoremediation

Phytoremediation techniques are suitable, aesthetically satisfying, and less alarming than the current techniques of physical and chemical process (Gerhardt et al. 2009). Applications of this technology are its efficiency in pollutant degradation, cheapness, and being suitable for wide diversity of contaminants, and on the whole it is an eco-friendly process. The major benefits of the phytotechnology through biomass accretion are its efficiency in decreasing heavy metal concentration to exceedingly lower limit and the use of cheap biosorbent materials (Wang and Chen 2009). Phytoremediation is currently enumerated as economical technology that can be used in the remediation of specific contaminated sites (Salamanca et al. 2015). Phytoremediation comprises numerous diverse schemes that result in contaminant degradation.

Phytoremediation is an economical substitute and rational strategy for cleaning ecosystem, mainly appropriate to enormous place that have moderately lower level of pollution (Padmavathiamma and Li 2007). This technology has been categorized as resourceful, productive alternative to the more renowned remediation techniques utilized at hazardous waste sites. Phytoremediation is considerably elite, economical explanation to several present inconveniences of soil pollution (LeDuc and Terry 2005). It is inexpensive (60–80 %) than expected physicochemical process, as it does not require expensive apparatus or exceptionally meticulous recruits. It is lucrative for enormous water reservoirs having lesser pollutant absorption and for expansive areas exhibiting lesser to considerably polluted surface soils (Conesa et al. 2012). It is significant to the widespread diversity of heavy metals and radionuclides and also important for remediating an extensive environmental pollutant such as organic and inorganic pollutants.

Phytoremediation is recognized current technique for the remediating of polluted soils, water, and ambient air (Eapen and D'souza 2005). Phytoremediation technique can offer reclamation of infertile soil polluted by excessive metal and salt concentration. It is relevant to diverse heavy metals and radionuclide least ecological interruption, degradation of soil and water pollutants, and civic approval (Succuro et al. 2009). Phytoextraction is known as an eco-friendly process to eliminate metals from contaminated soils in situ. This technique can be used in extensive areas and has been significant for heavy metal-polluted areas. It is an aesthetically agreeable, solar energy-derived remediation technology, and there is insignificant environmental interruption and in situ remediation conserve topsoil (Lin et al. 2008).

Field trials of phytoremediation reduce the range of soil deterioration in contrast to conventional techniques. It can be controlled with petite ecological disturbance with upper soil crust left in a utilizable state and may be cultivated for farming purpose. The organic contaminants can be detoxified CO_2 and H_2O, eradicating ecological toxicity (Rhodes 2013). Phytoremediation can be effectually replace other remediation technologies such as incineration, thermal vaporization, solvent washing, and soil washing techniques which are more extreme cleaning procedures since they disturb physiochemical and biological attributes of soil and form nonbiodegradable wastes (Jadia and Fulekar 2009). In contrast, phytoremediation rejeneuvates topsoil layer and leave no solid waste. Further advancement in phytoremediation is the release of a recyclable metal-rich plant filtrate.

5.9.1 Phytoremediation for Hydraulic Regulation of Pollutants

Plants function as hydraulic pumps with their roots drenched inside water table and establishing compact root mass that absorb enormous water. Poplar trees can lose 50–300 gallons of water per day through transpiration (Hoffman 2003). The water absorption via plants reduce the affinity of surface pollutants to trickle in drinking and ground water. The use of plants for instant water uptake is managed through hydraulic control. There are several advantages where plants are used for this function such as riparian corridors/buffer strips and vegetative caps (Isebrands et al. 2014).

5.9.2 Riparian Corridors

Riparian corridors (derived from "riparian" that signifies "placed on the bank of a river") or buffer strips are tools of phytoremediation that comprise of features of phytodegradation, phytovolatilization, and rhizodegradation to control, obstruct, or clean pollutants flowing in river or groundwater plume (Shukla et al. 2010). Uses of these processes avoid pollution from dispersing into surface and groundwater.

5.9.3 Vegetative Cover

Vegetative cover is a continuing, independent cap consisting of soil and plants increasing in and/or over waste in a landfill. This type of cover is an alternative to composite clay or plastic layer caps (Bortone et al. 2012). Plants control erosion and decline water seepage that might else permeate through landfill and structure polluted leachates (Rai 2008). Furthermore, a vegetative cap can be designed not only to control erosion and seepage of water but also to enhance the deterioration of chief materials in the landfill.

5.9.4 Phytoremediation to Treat Metal Contaminants

In order to clean the polluted locations, hyperaccumulators are employed to eliminate heavy metals from soil and groundwater through different processes like phytoextraction, rhizofiltration, and phytostabilization (Tangahu et al. 2011; Ali et al. 2013; Glick 2015).

5.9.5 Constructed Wetlands

World's famous smelter in alliance with Environment Canada employ constructed wetlands to remediate landfill leachate (Nivala et al. 2007; Madera-Parra et al. 2015). The most hazardous explosive found in groundwater was degraded through a constructed wetland planned and organized by Milan Ammunition Plant in Tennessee, USA (Best et al. 1999). In addition, bio-evaluation of ten aquatic plant species raised in growth chambers and hydroponic culture advanced to selection of specific crops like elodea (*Elodea Michx*), sago pondweed (*Stuckenia pectinatus*), and water grass (*Luziola fluitans*) (Qian et al. 1999).

5.9.6 Vegetative Caps

Vegetative caps are another alternative for remediation of landfill locations. Vegetative covers develop hydraulic pressure by amplifying the retention ability of soil, besides evaporation rates and transpiration modes of plants, thus declining water infiltration (Mills and Robinson 2003). They can facilitate degradation of wastes and contaminants within landfill. Zalesny et al.

(2006) reported the remediation of landfill in Slovenia using poplar trees to inspect build of leachate which might pollute ground and surface water.

5.9.7 Soil Conditioning

Evelin et al. (2012) illustrated reclamation potential of some plants which have crucial physiological methods and ability to accumulate substantial content of iron, manganese and magnesium, calcium, sodium, and chloride ions, thereby reducing overall salinity of the soil and associated effects on crop physiology and growth. The cultivation of salt-tolerant species like *Conocarpus erectus, Atriplex lentiformis* in saline water ensures high productivity even in arid regions (Padmavathiamma et al. 2014).

5.9.8 Fortification of Riparian Corridors

The plantation established for phytotechnologies provides supplementary advantage of fortification to crops in addition to remediation of contaminated soils (Gomes 2012). Licht and Isebrands (2005) reported that plants cultivated as a riparian corridor not only prevent agricultural pollution but also provide forest commodities such as nuts, fruits, lumber, or fiber. However, these crops don't translocate toxins from roots to shoots keeping harvest safe whereas roots remediate groundwater. Furthermore, polluted metropolitan sites might be cultivated to remediate soil and groundwater whereas simultaneously providing amusement.

5.10 Limitations of Phytoremediation Technology

Though phytoremediation has enormous advantages, there are various constraints that create hurdles for its extensive implementation. These disadvantages have been listed below.

Phytoremediation progress is governed by total biomass yield, root depth, soil chemistry, micrometeorology and extent of pollution, plant age, concentration, and effects of pollutant on plant life (Willey 2006).

Phytoremediation is a time-consuming technique as the plant takes several seasons to remediate contaminated sites (Dhillon and Dhillon 2015). Moreover, plant might accumulate cytotoxic compounds within tissues (Koptsik 2014).

The growth of plants lags the process of phytoremediation as extra time is requisite to phytoremediate contaminated site in contrast to traditional remediation technologies. Incineration cleans up polluted sites within weeks to months, whereas phytoextraction and other phytotechnologies necessitate for several years. Consequently, for intensely contaminated sites, phytoremediation might not be an ideal remediation process of assortment. However, phytoremediation can be recommended for sites where human contact is scarce or soil pollution require an instant response (Mendez and Maier 2008).

One of imprecise subject is replacement flanked by heavy element accumulation and productivity. In fact, tremendous harvestable dry-weight productivity of 10–20 $t^{-1}ha^{-1}y^{-1}$ would be anticipated, mainly for hyperaccumulators (Turpie et al. 2010). These significance for yield and heavy metal concentration should bind annual toxic element elimination competence regarding between 10 and 400 kg/ha/y, depending on the pollutant, plant species, and climatic factors (Nagajyoti et al. 2010; Wuana and Okieimen 2011). Root zone is a major constraint on phytoremediation process. For successful remediation process, plants require communion with the rhizospheric zone of plants. The rhizospheric zone must accommodate all pollutants, or plants must expand rhizospheric zone to contain all pollutants. Controlled to sites with lower pollution rate around rhizosphere of remediative plants, ground surface at the site ought to be modified to circumvent flooding or erosion (Sharma and Pandey 2014).

Root performance is significantly influenced by plant age. Generally, younger plant exhibits improved capacity to absorb ions than older plants irrespective of size. It is important to develop dynamic young plants for added capable plant exclusion. Enhanced levels of contaminants might decline plant growth and hence might restrict utilization on some sites (Tangahu et al. 2011). This phytotoxicity might advance toward phytoremediation scheme in which bulk waste is to switch with expensive ex situ processes that quickly decline thorough menace, while in situ phytoremediation is prolonged process that cleans contaminants from landfill areas.

The principal constraint in the phytoremediation of toxic metals is the highest level that can be amassed by plants (Surriya et al. 2014). Plants with higher concentrations of toxic metals, termed as "hyperaccumulators," generally exhibit concentration of about 2000 ppm (0.2 %) (Cd, Pb) to 2 % for less toxic (Zn, Ni, Cu) (Öztürk et al. 2016). Restricted to sites with lower pollutant percentage, the administration is generally limited to within range of soil surface, and groundwater within a few meters of the surface with soil modifications might be significant (Gavrilescu 2006). Some environmental pollution might occur every time plants are utilized related with pollutants from the soil. The providence of contaminant in huge biomass is intriguing subject. Although varied types of phytoremediation accommodate metal accumulation and require supervision of plants ingrained with metals, majority of plants do not accumulate significant content of organic contaminants. Although hyperaccumulator will require harvesting and also reusing or predisposing in compliance with suitable policies, generally phytoremediative plants do not require elimination strategy (Chehregani et al. 2009). Phytoextraction yields can be categorized as a detrimental waste; thus, elimination should be suitable. Exploitation of polluted plant biomass is a subjective issue pollutant that might enter food chain through animals/insects that eat metal-polluted plants (Pathak et al. 2012). The growth of plants is hindered by climatic and hydrologic surroundings. Introduction of exotic species might implicate biodiversity (Butcher 2009; Kühl et al. 2010). Phytoremediation technology is the most feasible method to clean heavy metal-polluted environment.

There are few limitations to execute phytoremediation using biodiversity (Åslund and Zeeb 2010). To significant stage, these comprise extensive pollution over vegetation cover and food chain and are remarkably complex to launch and maintain foliage on polluted sites, for example, mine tailings with high content of remaining metals (Borker et al. 2013). For metal pollutants, plants exhibit the prospective for phytoextraction and revival of pollutants into shoots (Gavrilescu et al. 2015), rhizofiltration, or stabilizing landfills through erosion control and evapotranspiration of huge water reservoirs (phytostabilization) (Mukhopadhyay and Maiti 2010; Adelana et al. 2011).

Phytoextraction has major benefit since the plant biomass retaining the pollutant can be a preserved. For example, (i) biomass that possess selenium (Se), an essential nutrient, has been translocated to vegetative portion where Se is deficient and hence utilized for animal feed (Pilon-Smits and Quinn 2010), and (ii) metal hyperaccumulators are specifically involved in biogeochemical prospecting of minerals (Prasad and de Oliveira Freitas 2003).

Rhizofiltration limitations:

(i) The pH of the soil sap need to be regularly adjusted to obtain optimal metal uptake.
(ii) The chemical speciation and communication of all species in the influent have to be implicit for suitable role.
(iii) The properly organized scheme is necessary to supervise influent content and flow rate.
(iv) Terrestrial plants need to be raised in a greenhouse or nursery and then located in the rhizofiltration system. (v) Cyclic production and plant dumping are important.
(v) Laboratory- and greenhouse-raised plants exhibiting high metal immobility and uptake exhibit unsatisfactory response in field trail.

Phytovolatilization limitations: (i) the contaminant like Se is released into the atmosphere (Chan et al. 2014). As a result, adequate understanding is necessary for phytoremediation-based technique integrated with the environment, e.g., greenbelts (valuable ecological niches, primarily in metropolitan developed locale) constructed wetlands in which *Eichhornia crassipes* (water hyacinth), *Hydrocotyle umbellata* (pennywort), *Lemna minor* (duckweed), and *Azolla pinnata* (water velvet) are conserved and managed which can absorb Pb, Cu, Cd, Fe, and Hg from aqueous solutions (Brundu et al. 2012; Ensley et al. 2012).

5.10.1 Relevance of Phytoremediation

5.10.1.1 Constructed Wetlands

Developed countries like Europe employ extensively constructed wetlands for the remediation of polluted sites. Several paradigms have been demonstrated in varied meetings. European commission has maintained the implementation of a 10,000 m^2 wetland in Portugal, to remediate industrial effluents having aniline, nitrobenzene, and sulfanilic acid (Vymazal 2014).

5.10.1.2 Short Rotation Coppice Forestry

This refers to practice of cultivating trees, poplars, or willows raised for less than 15 years and harvested to get products like paper and pulp (Aylott et al. 2008). Exclusively, coppicing comprises of cutting the trunk from the base at intervals of 2–3 years, and shoots appear from the stump. This kind of forestry also indicates a reserve of renewable energy, consisting at the same time a sink for atmospheric carbon. Utilization of plants which can accumulate heavy metals consumes CO_2 and generates biomass and binds forestry with phytotechnologies (Xue et al. 2015). Several researchers are subsequently investigating metal uptake in willow and poplar, so as to calculate the biodiversity existing among cultivars, clones, and accessions. As a present concept, poplar clones were analyzed for numerous heavy metal uptake, selecting how Cd, Zn, and Al were absorbed with increased competency (Dessaux et al. 2016).

5.10.1.3 Communications with Microbiome

It is recognized that plant–microbe communications hold crucial role in phytoremediation. Current research finding of the role of endophytic bacteria in phytoremediation has led to several interesting negotiations (Newman and Reynolds 2005; Ryan et al. 2008; Ma et al. 2011; Weyens et al. 2015). Barac et al. (2004) reported reduced phytotoxicity and enhanced degradation of toluene in yellow lupine plants by engineering endophytic bacteria of the species *Burkholderia cepacia* with plasmid pTOM.

5.10.1.4 Atmospheric Contaminants

Phytotechnologies have traditionally been limited to pollutants accessible via plant roots, either in soil and sediments or in water. However, pollutants might infiltrate the plant via atmosphere, and a novel relevance of the phytotechnologies might be removal of pollutants from the troposphere (Guerra et al. 2015). Nitrogen dioxide is obnoxious pollutant which is absorbed through stomata and incorporated into organic compounds (Kvesitadze et al. 2006). There is an extensive discrepancy among plant taxa in this issue, and an assessment of apropos 300 species revealed that the mainly capable plant is *Eucalyptus viminalis*, 657 times extra competent than *Tillandsia*, the less competent taxon (Marmiroli et al. 2006). These plants could be utilized to congregate "green walls," layering the vertical surfaces of building where plants are able to integrate NO_2 in huge quantities. Jin et al. (2009) has illustrated an assenting consequence of NO_2 on plant biomass growth, defining it as a "plant vitalization."

5.10.1.5 Natural Remediation

The current aptitude to phytotechnologies is the pioneering enterprise convene by the purported assisted natural remediation, perceptibly contemporary apparatus (Adriano et al. 2004). In an assisted natural remediation, modifications are made to the soil to assist natural processes of remediation. In the case of metals, these amendments improve immobilization together with complexation, adsorption, precipitation, and chemical reactions. The main justification is to lower the bioavailability of the metal and not its total absorption (Pérez-de-Mora et al. 2006).

5.10.1.6 Biofortification

Pervasive investigation on the boundary amid plant tissues, heavy metals, and/or trace elements has led to the discernment of biofortification, in which plants amplified in micronutrient concentration are termed as assistance in addition to malnutrition (Zhao and McGrath 2009). Segregated from phytoaccumulation of metals, which is known as a menace for the food chain, biofortification of crops with overt elements may become valuable (Zhu et al. 2009). Fortified crops are suitable for growth on micronutrient-poor soil since their bioconcentration competency will escort to high concentration of micronutrients in edible tissues. Perceptive of method controlling metal accretion is a requirement for elucidating the biochemical source of these events. Statistics is also necessary for those antinutrients that reduce heavy metal convenience, e.g., phytic acid, fibers, and polyphenols (Yu et al. 2012).

5.10.1.7 Glucosinolates and Biofumigation

Utilization of plant-based chemicals is a recent technology for pest control in agricultural sector. In particular, several Brassicaceae are conquered in the resistance against biotic stress in agriculture owing to the release of specific secondary metabolites known as glucosinolates (Björkman et al. 2011). Glucosinolates are sulfur-containing compounds that exhibit anticarcinogenic activity supplying characteristic basis of cruciferous plants, and their degradation modules can avert herbivores and impede microbes (Gimsing and Kirkegaard 2009). In this context, it has been recommended to use Brassicaceae as "green manure" to be amplified to the soil through soil conditioning, to facilitate reduction in pathogenic attack (Kruger et al. 2015). Biofumigation thus exist as significant substitute for the application of chemical fumigants. The most exhilarating aspect to be investigated is the potential relationship between the production of glucosinolates in the plant and the incidence of heavy metals in the environment which function as inducers or repressors (Dar et al. 2015). Since glucosinolates possess sulfur-containing complexes such as metallothioneins and phytochelatins, they might affect sulfur metabolism of heavy metal sequestrating peptide application (Asgher et al. 2015). For natural process of remediation particularly in case of heavy metals, alterations to soil are being reinforced to induce immobilization, complexation, adsorption, precipitation, and chemical reactions. The principal objective behind this alteration is to lower bioavailability of the metals (Szczygłowska et al. 2011).

5.10.1.8 Function of Biodiversity in Phytoremediation

The relevance of biodiversity that exist across the globe is exceedingly known for its remediation of metal-contaminated and metal-polluted ecosystems. This subject is promising as a drastic spot of research advance with utmost significance in the contemporary area of environmental biotechnology (Denton 2007). Several microbes comprising mycorrhizal and non-mycorrhizal fungi, agricultural and vegetable crops, ornamentals, and wild metal hyperaccumulating plants are investigated both in situ and ex situ for remediating metalliferous areas. There are approximately 400 plants that hyperaccumulate metals. Most of these plants belong to families Asteraceae, Brassicaceae, Caryophyllaceae, Cyperaceae, Cunoniaceae, Fabaceae, Flacourtiaceae, Lamiaceae, Poaceae, Violaceae, and Euphobiaceae (Smolińska and Król-Domańska 2010). Most of hyperaccumulators belong to family to counting about 11 genera and 87 species. Diverse genera of Brassicaceae are recognized to accrue metals. Ni hyperaccumulation is meticulous in 7 genera and 72 species and Zn in 3 genera and 20 species (Peer et al. 2006). *Thlaspi* species are known to hyperaccumulate numerous metals, i.e., *T. caerulescens* = Cd, Ni, Pb, and Zn; *T. goesingense* = Ni and Zn; *T. ochroleucum* = Ni and Zn; and *T. rotundifolium* = Ni, Pb, and Zn (Koptsik 2014). Plants that hyperaccumulate metals have significant potential for relevance in heavy metal cleanup in the environment. Significant progress in phytoremediation is established metals and radionuclides (Jadia and Fulekar 2009). These practices involve rising of plants hydroponically and transplanting them into metal-contaminated waters where plants accumulate plant roots and shoots. Plants once saturated with heavy metals, roots, or whole plants are harvested for elimination. Research however recommends accumulation of heavy metals in plant roots. Several aquatic species have the capability for heavy metal cleanup, e.g., water hyacinth (*Eichhornia crassipes* (Mart.) Solms), pennywort (*Hydrocotyle umbellata* L.), and duckweed (*Lemna minor* L.) (Khilji 2008). Elimination of Cd, Cr, Cu, Ni, Pb, and Zn through the accumulation of said elements in roots of *Brassica* species and sunflower is known to eliminate Pb, U, ^{137}Cs, and ^{90}Sr from hydroponic solutions (Chandra 2015). Hence, freshwater, marine, and estuarine dwelling aquatic plants function as vessel for several metals. Hyperaccumulators accumulate significant amount of metal in their tissue in spite of the metal accumulation in the soil, provided that the metal to be explored is present (Rascio and Navari-Izzo 2011). The phytoextraction method involves the use of plants to aid the elimination of metal contamination from a soil stratum. Actually, hyperaccumulators are raised and transplanted in metal-contaminated soils via intensive cultivation practices. In case heavy metals are present in scanty amount, diverse chelates or acidifying agents would be employed to discharge them into soil sap. Polat et al. (2004) employed ammonium thiocyanate and natural zeolites as soil conditioners to modulate the soil matrix. Several artificial cross-linked polyacrylates and hydrogels have isolated heavy metals from plant roots and thus hampered heavy metal toxicity in roots (Hüttermann et al. 2009). To ensure elimination, heavy metal is bioavailable so as to be absorbed via plant roots. For these purposes, different chemicals such as acidifying fertilizer salts and chelating complexes have been suggested to increase bioavailability. Park et al. (2011) reported the higher incidence of heavy metal bioavailability near soil organic matter at low pH. Phytoextraction process has enhanced the availability of heavy metal by augmenting soil with acidifying agents.

Evangelou et al. (2007) illustrated the potential of chelates to enhance the phytoextraction of several metal pollutants comprising Cd, Cu, Ni, Pb, and Zn.

References

Adelana SO, Adeosun T, Adesina A, Ojuroye M (2011) Environmental pollution and remediation: challenges and management of oil Spillage in the Nigerian coastal areas. Am J Sci Ind Res 2(6):834–845

Adriano DC, Wenzel WW, Vangronsveld J, Bolan NS (2004) Role of assisted natural remediation in environmental cleanup. Geoderma 122(2):121–142

Ali H, Khan E, Sajad MA (2013) Phytoremediation of heavy metals—concepts and applications. Chemosphere 91(7):869–881

Alkorta I, Hernández-Allica J, Becerril JM, Amezaga I, Albizu I, Onaindia M, Garbisu C (2004) Chelate-enhanced phytoremediation of soils polluted with heavy metals. Rev Environ Sci Biotechnol 3(1):55–70

Ansari MKA, Shao HB, Umar S, Ahmad A, Ansari SH, Iqbal M, Owens G (2013) Screening Indian mustard genotypes for phytoremediating arsenic-contaminated soils. Clean Soil Air Water 41(2):195–201

Antosiewicz DM, Barabasz A, Siemianowski O (2013) Phenotypic and molecular consequences of overexpression of metal-homeostasis genes. Front Plant Sci 5:80–80

Asgher M, Khan MIR, Fatma M, Khan NA, Chakraborty U, Chakraborty B (2015) Potentiality of ethylene in sulfur-mediated counteracting adverse effects of cadmium in plants. In: Abiotic stresses in crop plants, vol 136. CAB International, Boston

Åslund MW, Zeeb BA (2010) A review of recent research developments into the potential for phytoextraction of persistent organic pollutants (POPs) from weathered, contaminated soil. In: Application of phytotechnologies for cleanup of industrial, agricultural, and wastewater contamination. Springer, Dordrecht, pp 35–59

Aylott MJ, Casella E, Tubby I, Street NR, Smith P, Taylor G (2008) Yield and spatial supply of bioenergy poplar and willow short-rotation coppice in the UK. New Phytol 178(2):358–370

Babaeian E, Homaee M, Rahnemaie R (2016) Chelate-enhanced phytoextraction and phytostabilization of lead-contaminated soils by carrot (Daucus carota). Arch Agron Soil Sci 62(3):339–358

Barac T, Taghavi S, Borremans B, Provoost A, Oeyen L, Colpaert JV, … van der Lelie D (2004) Engineered endophytic bacteria improve phytoremediation of water-soluble, volatile, organic pollutants. Nat Biotechnol 22(5):583–588

Bardgett RD, Mommer L, De Vries FT (2014) Going underground: root traits as drivers of ecosystem processes. Trends Ecol Evol 29(12):692–699

Bernard C, Roosens N, Czernic P, Lebrun M, Verbruggen N (2004) A novel CPx-ATPase from the cadmium hyperaccumulator Thlaspi caerulescens. FEBS Lett 569(1):140–148

Best EP, Sprecher SL, Larson SL, Fredrickson HL, Bader DF (1999) Environmental behavior of explosives in groundwater from the Milan Army Ammunition Plant in aquatic and wetland plant treatments. Removal, mass balances and fate in groundwater of TNT and RDX. Chemosphere 38(14):3383–96

Björkman M, Klingen I, Birch AN, Bones A.M, Bruce TJ, Johansen TJ, … Stewart D (2011) Phytochemicals of Brassicaceae in plant protection and human health–influences of climate, environment and agronomic practice. Phytochemistry 72(7):538–556

Bolan N, Kunhikrishnan A, Thangarajan R, Kumpiene J, Park J, Makino T, Scheckel K (2014) Remediation of heavy metal (loid) s contaminated soils–to mobilize or to immobilize? J Hazard Mater 266:141–166

Borker AR, Mane AV, Saratale GD, Pathade GR (2013) Phytoremediation potential of Eichhornia crassipes for the treatment of cadmium in relation with biochemical and water parameters. Emirates J Food Agric 25(6):443

Bortone I, Chianese S, Di Nardo A, Di Natale M, Erto A, Musmarra D (2012) Groundwater protection by permeable adsorbing barriers at solid waste landfills. In: Landfills: waste management, regional practices and environmental impact. Nova Science Publishers, Inc., Hauppauge, ISBN 978-1-62257-121, 5

Brundu G, Stinca A, Angius L, Bonanomi G, Celesti-Grapow L, D'Auria G, … Spigno P (2012) Pistia stratiotes L. and Eichhornia crassipes (Mart.) Solms.: emerging invasive alien hydrophytes in Campania and Sardinia (Italy). EPPO Bull 42(3):568–579

Buescher E, Achberger T, Amusan I, Giannini A, Ochsenfeld C, Rus A, … Guerinot ML (2010) Natural genetic variation in selected populations of Arabidopsis thaliana is associated with ionomic differences. PLoS One 5(6):e11081

Butcher DJ (2009) Phytoremediation of lead in soil: recent applications and future prospects. Appl Spectrosc Rev 44(2):123–139

Chan BKC, Luo T, Wang RY (2014, September) Phytoremediation of heavy metal contaminated soil by selection of plant species. In Legislation, technology and practice of mine land reclamation: proceedings of the Beijing international symposium on land reclamation and ecological restoration (LRER 2014), Beijing, China, 16–19 October 2014. CRC Press, pp 429

Chandra R (ed) (2015) Advances in biodegradation and bioremediation of industrial waste. CRC Press, Boca Raton

Chehregani A, Noori M, Yazdi HL (2009) Phytoremediation of heavy-metal-polluted soils: screening for new accumulator plants in Angouran mine (Iran) and evaluation of removal ability. Ecotoxicol Environ Saf 72(5):1349–1353

Chen H, Cutright TJ (2002) The interactive effects of chelator, fertilizer, and rhizobacteria for enhancing phytoremediation of heavy metal contaminated soil. J Soils Sediments 2(4):203–210

Conesa HM, Evangelou MW, Robinson BH, Schulin R (2012) A critical view of current state of phytotechnologies to remediate soils: still a promising tool? Sci World J 2012:1–10

Cosio C, Martinoia E, Keller C (2004) Hyperaccumulation of cadmium and zinc in Thlaspi caerulescens and Arabidopsis halleri at the leaf cellular level. Plant Physiol 134(2):716–725

Craciun AR, Courbot M, Bourgis F, Salis P, Saumitou-Laprade P, Verbruggen N (2006) Comparative cDNA-AFLP analysis of Cd-tolerant and-sensitive genotypes derived from crosses between the Cd hyperaccumulator Arabidopsis halleri and Arabidopsis lyrata ssp. petraea. J Exp Bot 57(12):2967–2983

Daniell H (2002) Molecular strategies for gene containment in transgenic crops. Nat Biotechnol 20(6):581–586

Dar TA, Uddin M, Khan MMA, Hakeem KR, Jaleel H (2015) Jasmonates counter plant stress: a review. Environ Exp Bot 115:49–57

Dauchy S, Dutheil F, Weaver RJ, Chassoux F, Daumas-Duport C, Couraud PO, … Declèves X (2008) ABC transporters, cytochromes P450 and their main transcription factors: expression at the human blood–brain barrier. J Neurochem 107(6):1518–1528

de Mello-Farias PC, Chaves ALS, Lencina CL (2011) Transgenic plants for enhanced phytoremediation-physiological studies. INTECH Open Access Publisher Rijeka, Croatia, EU

Denton BP (2007) Advances in phytoremediation of heavy metals using plant growth promoting bacteria and fungi. Warning: get_class () expects parameter 1 to be object, array given in/home/vhosts/ejournal/user-dir/htdocs/classes/cache/GenericCache. inc. php on line 63 MMG 445. Basic Biotechnol eJ 3(1):1–5

Dessaux Y, Grandclément C, Faure D (2016) Engineering the rhizosphere. Trends in plant science 21(3):266–78

Dhillon KS, Dhillon SK (2015) Phytoremediation of selenium contaminated soils: strategies and limitations. In: Global advances in selenium research from theory to application: proceedings of the 4th international conference on selenium in the environment and human health 2015. CRC Press, pp 201

Didierjean L, Gondet L, Perkins R, Lau SMC, Schaller H, O'Keefe DP, Werck-Reichhart D (2002) Engineering herbicide metabolism in tobacco and Arabidopsis with CYP76B1, a cytochrome P450 enzyme from Jerusalem artichoke. Plant Physiol 130(1):179–189

Donner E, Punshon T, Guerinot ML, Lombi E (2012) Functional characterisation of metal (loid) processes in planta through the integration of synchrotron techniques and plant molecular biology. Anal Bioanal Chem 402(10):3287–3298

Eapen S, D'souza SF (2005) Prospects of genetic engineering of plants for phytoremediation of toxic metals. Biotechnol Adv 23(2):97–114

Emongor V (2007) Biosorption of lead from aqueous solutions of varied pH by kale plants (Brasicca oleraceae var. acephala). J Agric Food Environ Sci 1(2):90–91

Ensley BD, Raskin I, Salt DE (2012) Phytoremediation applications for removing heavy metal contamination from soil and water. Biotechnol Sustain Environ 54:59

Escobar MP, Dussán J (2016) Phytoremediation potential of chromium and lead by Alnus acuminata subsp. acuminata. Environmental progress & sustainable energy 35(4):942–948

Evangelou MW, Ebel M, Schaeffer A (2007) Chelate assisted phytoextraction of heavy metals from soil. Effect, mechanism, toxicity, and fate of chelating agents. Chemosphere 68(6):989–1003

Evelin H, Giri B, Kapoor R (2012) Contribution of Glomus intraradices inoculation to nutrient acquisition and mitigation of ionic imbalance in NaCl-stressed Trigonella foenum-graecum. Mycorrhiza 22(3):203–217

Gavrilescu M (2006) Overview of in situ remediation technologies for sites and groundwater. Environ Eng & Manag J (EEMJ) 5(1)

Gavrilescu M, Demnerová K, Aamand J, Agathos S, Fava F (2015) Emerging pollutants in the environment: present and future challenges in biomonitoring, ecological risks and bioremediation. New Biotechnol 32(1):147–156

Gerhardt KE, Huang XD, Glick BR, Greenberg BM (2009) Phytoremediation and rhizoremediation of organic soil contaminants: potential and challenges. Plant Sci 176(1):20–30

Gimsing AL, Kirkegaard JA (2009) Glucosinolates and biofumigation: fate of glucosinolates and their hydrolysis products in soil. Phytochem Rev 8(1):299–310

Glick BR (2015) Phytoremediation. In: Beneficial plant-bacterial interactions. Springer International Publishing, Cham, pp 191–221

Gomes HI (2012) Phytoremediation for bioenergy: challenges and opportunities. Environ Technol Rev 1(1):59–66

Guerra FP, Reyes L, Vergara-Jaque A, Campos-Hernández C, Gutiérrez A, Pérez-Díaz J, … Ruíz-Lara S (2015) Populus deltoides Kunitz trypsin inhibitor 3 confers metal tolerance and binds copper, revealing a new defensive role against heavy metal stress. Environ Exp Bot 115:28–37

Hoffman KM (2003) Phytoremediation of hydrocarbons by Arroyo Willows in controlled laboratory microcosms. Doctoral dissertation, California Polytechnic State University San Luis Obispo

Hüttermann A, Orikiriza LJ, Agaba H (2009) Application of superabsorbent polymers for improving the ecological chemistry of degraded or polluted lands. Clean Soil Air Water 37(7):517–526

Isebrands JG, Aronsson P, Carlson M, Ceulemans R, Coleman M, Dickinson N, … Johnson JD (2014) 6 environmental applications of Poplars and Willows. In: Poplars and Willows: trees for society and the environment. CABI Publishing Wallingford, pp 258–336

Jabeen R, Ahmad A, Iqbal M (2009) Phytoremediation of heavy metals: physiological and molecular mechanisms. Bot Rev 75(4):339–364

Jadia CD, Fulekar MH (2009) Phytoremediation of heavy metals: recent techniques. Afr J Biotechnol 8(6):491–499

Jain M, Ghanashyam C, Bhattacharjee A (2010) Comprehensive expression analysis suggests overlapping and specific roles of rice glutathione S-transferase genes during development and stress responses. BMC Genomics 11(1):73

Jin CW, Du ST, Zhang YS, Tang C, Lin XY (2009) Atmospheric nitric oxide stimulates plant growth and improves the quality of spinach (Spinacia oleracea). Ann Appl Biol 155(1):113–120

Khilji S (2008) Rhizofiltration of heavy metals from the tannery sludge by the anchored hydrophyte, Hydrocotyle umbellata L. Afr J Biotechnol 7(20):3714–3720

Kirchner S (2009) The AtIREGs-characterization of a new family of metal transporters in Arabidopsis thaliana. Thesis submitted to Faculty of Agricultural Sciences at the University of Hohenheim. pp 235

Koptsik GN (2014) Problems and prospects concerning the phytoremediation of heavy metal polluted soils: a review. Eurasian Soil Sci 47(9):923–939

Kruger DHM, Fourie JC, Malan AP (2015) Control potential of Brassicaceae cover crops as green manure and their host status for Meloidogyne javanica and Criconemoides xenoplax. S Afr J Enol Vitic 36(1):165–174

Kühl N, Moschen R, Wagner S, Brewer S, Peyron O (2010) A multiproxy record of late Holocene natural and anthropogenic environmental change from the Sphagnum peat bog Dürres Maar, Germany: implications for quantitative climate reconstructions based on pollen. J Quat Sci 25(5):675–688

Kvesitadze G, Khatisashvili G, Sadunishvili T, Ramsden JJ (2006) Biochemical mechanisms of detoxification in higher plants: basis of phytoremediation. Springer Science & Business Media, Berlin/Heidelberg

LeDuc DL, Terry N (2005) Phytoremediation of toxic trace elements in soil and water. J Ind Microbiol Biotechnol 32(11–12):514–520

Leoni C, Volpicella M, De Leo F, Gallerani R, Ceci LR (2011) Genome walking in eukaryotes. FEBS J 278(21):3953–3977

Leštan D, Luo CL, Li XD (2008) The use of chelating agents in the remediation of metal-contaminated soils: a review. Environ Pollut 153(1):3–13

Licht LA, Isebrands JG (2005) Linking phytoremediated pollutant removal to biomass economic opportunities. Biomass Bioenergy 28(2):203–218

Lin C, Negev I, Eshel G, Banin A (2008) In situ accumulation of copper, chromium, nickel, and zinc in soils used for long-term waste water reclamation. J Environ Qual 37(4):1477–1487

Lu X, Zhang T, Fang HP (2011) Bacteria-mediated PAH degradation in soil and sediment. Appl Microbiol Biotechnol 89:1357–1371

Ma Y, Prasad MNV, Rajkumar M, Freitas H (2011) Plant growth promoting rhizobacteria and endophytes accelerate phytoremediation of metalliferous soils. Biotechnol Adv 29(2):248–258

Madera-Parra CA, Peña-Salamanca EJ, Peña MR, Rousseau DPL, Lens PNL (2015) Phytoremediation of landfill leachate with Colocasia esculenta, Gynerum sagittatum and Heliconia psittacorum in constructed wetlands. Int J Phytoremediation 17(1):16–24

Maier RM, Pepper IL, Gerba CP (2009) Environmental microbiology, vol 397, 2nd edn. Methods in enzymology. ISSN 0076-6879. Academic, New York

Marmiroli M, Gonnelli C, Maestri E, Gabbrielli R, Marmiroli N (2004) Localisation of nickel and mineral nutrients Ca, K, Fe, Mg by Scanning Electron Microscopy microanalysis in tissues of the nickel-hyperaccumulator Alyssum bertolonii Desv. and the non-accumulator Alyssum montanum L. *Plant*. Biosystems 138(3):231–243

Marmiroli M, Antonioli G, Maestri E, Marmiroli N (2005) Evidence of the involvement of plant ligno-cellulosic structure in the sequestration of Pb: an X-ray spectroscopy-based analysis. Environ Pollut 134(2):217–227

Marmiroli N, Marmiroli M, Maestri E (2006) Phytoremediation and phytotechnologies: a review for the present and the future. In: Soil and water pollution monitoring, protection and remediation. Springer, Dordrecht, pp 403–416

McGrath SP, Zhao FJ (2003) Phytoextraction of metals and metalloids from contaminated soils. Curr Opin Biotechnol 14(3):277–282

Mendez MO, Maier RM (2008) Phytoremediation of mine tailings in temperate and arid environments. Rev Environ Science Bio/Technol 7(1):47–59

Merkl N, Schultze-Kraft R, Infante C (2005) Phytoremediation in the tropics–influence of heavy crude oil on root morphological characteristics of graminoids. Environ Pollut 138(1):86–91

Mertens J, Vervaeke P, De Schrijver A, Luyssaert S (2004) Metal uptake by young trees from dredged brackish sediment: limitations and possibilities for phytoextraction and phytostabilisation. Sci Total Environ 326(1):209–215

Mills TM, Robinson BH (2003) Hydrologic management of contaminated sites using vegetation. Encyclopedia of Water Science 1 DOI: 10.1081/E-EWS 120020436, Marcel Decker INC

Mukhopadhyay S, Maiti SK (2010) Phytoremediation of metal mine waste. Appl Ecol Environ Res 8(3):207–222

Nagajyoti PC, Lee KD, Sreekanth TVM (2010) Heavy metals, occurrence and toxicity for plants: a review. Environ Chem Lett 8(3):199–216

Newman LA, Reynolds CM (2005) Bacteria and phytoremediation: new uses for endophytic bacteria in plants. Trends Biotechnol 23(1):6–8

Nivala J, Hoos MB, Cross C, Wallace S, Parkin G (2007) Treatment of landfill leachate using an aerated, horizontal subsurface-flow constructed wetland. Sci Total Environ 380(1):19–27

Nowack B, Schulin R, Robinson BH (2006) Critical assessment of chelant-enhanced metal phytoextraction. Environ Sci Technol 40(17):5225–5232

Öztürk M, Ashraf M, Aksoy A, Ahmad MSA, Hakeem KR (2016) Plants, pollutants and remediation. Springer, Dordrecht

Padmavathiamma PK, Li LY (2007) Phytoremediation technology: hyper-accumulation metals in plants. Water Air Soil Pollut 184(1–4):105–126

Padmavathiamma PK, Ahmed M, Rahman HA (2014) Phytoremediation-A sustainable approach for contaminant remediation in arid and semi-arid regions-a review. Emirates J Food Agric 26(9):757

Pantola RC, Alam A (2014) Potential of brassicaceae burnett (Mustard family; Angiosperms) in phytoremediation of heavy metals. Int J Sci Res Environ Sci 2(4):120

Park JH, Lamb D, Paneerselvam P, Choppala G, Bolan N, Chung JW (2011) Role of organic amendments on enhanced bioremediation of heavy metal (loid) contaminated soils. J Hazard Mater 185(2):549–574

Parmar S, Singh V (2015) Phytoremediation approaches for heavy metal pollution: a review. J Plant Sci Res 2(2):135

Pathak AK, Singh M, Kumar V, TrivedI A (2012) Phytoremediation of municipal solid waste landfill site: a review. J Chem Chem Sci 2(1):1–92

Payne KA, Bowen HC, Hammond JP, Hampton CR, Lynn JR, Mead A, … Broadley MR (2004) Natural genetic variation in caesium (Cs) accumulation by Arabidopsis thaliana. New Phytol 162(2):535–548

Peer WA, Mahmoudian M, Freeman JL, Lahner B, Richards EL, Reeves RD, … Salt DE (2006) Assessment of plants from the Brassicaceae family as genetic models for the study of nickel and zinc hyperaccumulation. New Phytol 172(2):248–260

Pérez-de-Mora A, Madejón E, Burgos P, Cabrera F (2006) Trace element availability and plant growth in a mine-spill contaminated soil under assisted natural remediation I. Soils. Sci Total Environ 363(1):28–37

Perfus-Barbeoch L, Leonhardt N, Vavasseur A, Forestier C (2002) Heavy metal toxicity: cadmium permeates through calcium channels and disturbs the plant water status. Plant J 32:539–548

Persans MW, Nieman K, Salt DE (2001) Functional activity and role of cation-efflux family members in Ni hyperaccumulation in Thlaspi goesingense. Proc Nat Acad Sci USA 98:9995–10000. Pittman JK

Pilon-Smits EA, Quinn CF (2010) Selenium metabolism in plants. In: Cell biology of metals and nutrients. Springer, Berlin/Heidelberg, pp 225–241

Polat E, Karaca M, Demir H, Naci-Onus A (2004) Use of natural zeolite (clinoptilolite) in agriculture. J Fruit Ornamental Plant Res 12(1):183–189

Prasad MNV (2013) Metallophytes – Properties, Functions and Applications. The Botanica 62 & 63: 17–26

Qian JH, Zayed A, Zhu YL, Yu M, Terry N (1999) Phytoaccumulation of trace elements by wetland plants: III. Uptake and accumulation of ten trace elements by twelve plant species. J Environ Qual 28(5):1448–1455

Rai PK (2008) Heavy metal pollution in aquatic ecosystems and its phytoremediation using wetland plants:

an ecosustainable approach. Int J Phytorem 10(2):133–160
Rajkumar M, Ae N, Prasad MNV, Freitas H (2010) Potential of siderophore-producing bacteria for improving heavy metal phytoextraction. Trend Biotechnol 28(3):142–149
Rascio N, Navari-Izzo F (2011) Heavy metal hyperaccumulating plants: how and why do they do it? And what makes them so interesting? Plant Sci 180(2):169–181
Rawat K, Fulekar MH, Pathak B (2012) Rhizofiltration: a green technology for remediation of heavy metals. Intl J Inno Biosci 2(4):193–199
Rhodes CJ (2013) Applications of bioremediation and phytoremediation. Sci Prog 96(4):417–427
Rubin GM, Yandell MD, Wortman JR, Gabor GL, Nelson CR, Hariharan IK, … Cherry JM (2000) Comparative genomics of the eukaryotes. Science 287(5461):2204–2215
Ryan RP, Germaine K, Franks A, Ryan DJ, Dowling DN (2008) Bacterial endophytes: recent developments and applications. FEMS Microbiol Lett 278(1):1–9
Sakakibara M, Watanabe A, Inoue M, Sano S, Kaise T (2010) Phytoextraction and phytovolatilization of arsenic from As-contaminated soils by Pteris vittata. Proc Ann Int Conf Soil Sediment Water Energy 12(1):26
Salamanca EJP, Madera-Parra CA, Avila-Williams CA, Rengifo-Gallego AL, Ríos DA (2015) Phytoremediation using terrestrial plants. In: Phytoremediation. Springer International Publishing, Berlin/New York, pp 305–319
Santiago DL, Maiti RK, Pinero JLH, Oreja JAG (2004) Plant based bioremediation and mechanisms of heavy metal tolerance of plants: a review. Proc Indian Nat Sci Acad B 70:1–12
Sappl PG, Onate-Sanchez L, Singh KB, Millar AH (2004) Proteomic analysis of glutathione S-transferases of Arabidopsis thaliana reveals differential salicylic acid-induced expression of the plant-specific phi and tau classes. Plant Mol Biol 54(2):205–219
Sarret G, Saumitou-Laprade P, Bert V, Proux O, Hazemann JL, Traverse A, … & Manceau A (2002) Forms of zinc accumulated in the hyperaccumulatorArabidopsis halleri. Plant Physiol 130(4):1815–1826
Sasaki A, Yamaji N, Ma JF (2014) Overexpression of OsHMA3 enhances Cd tolerance and expression of Zn transporter genes in rice. J Exp Botany 65(20):6013–6021
Sharma P, Pandey S (2014) Status of phytoremediation in world scenario. Int J Environ Bioremed Biodegrad 2(4):178–191
Sheoran V, Sheoran AS, Poonia P (2012) Phytoremediation technologies. Environ Res J 393:168–214
Shukla KP, Singh NK, Sharma S (2010) Bioremediation: developments, current practices and perspectives. Genet Eng Biotechnol J 3:1–20
Singh D, Kashyap A, Pandey RV, Saini KS (2011) Novel advances in cytochrome P450 research. Drug Discov Today 16(17):793–799
Sinha RK, Valani D, Sinha S, Singh S, Herat S (2010) Bioremediation of contaminated sites: a low-cost nature's biotechnology for environmental clean up by versatile microbes, plants & earthworms. In: Timo F, Johann H (eds) Solid waste management and environmental remediation. Nova Science Publishers, Inc, New York, ISBN: 978-1-60741-761-3
Smolińska B, Król-Domańska K (2010) Phytoextraction of heavy metal contaminated soils. Scientific Papers Technical University of Lodz 1081–1074:89–98
Soranzo N, Gorla MS, Mizzi L, De Toma G, Frova C (2004) Organisation and structural evolution of the rice glutathione S-transferase gene family. Mol Gen Genomics 271(5):511–521
Soudek P, Tykva R, Vaněk T (2004) Laboratory analyses of 137 Cs uptake by sunflower, reed and poplar. Chemosphere 55(7):1081–1087
Su ZH, Xu ZS, Peng RH, Tian YS, Zhao W, Han HJ, … Wu AZ (2012) Phytoremediation of trichlorophenol by phase II metabolism in transgenic Arabidopsis overexpressing a Populus glucosyltransferase. Environ Sci Technol 46(7):4016–4024
Succuro JS, McDonald SS, Lu CR (2009) Phytoremediation: the wave of the future. In: Recent advances in plant biotechnology. Springer, New York, pp 119–135
Surriya O, Sarah SS, Waqar K, Gul Kazi A (2014) Phytoremediation of soils: prospects and challenges. Academic Press, London, pp 1–1
Szczygłowska M, Piekarska A, Konieczka P, Namieśnik J (2011) Use of brassica plants in the phytoremediation and biofumigation processes. Int J Mol Sci 12(11):7760–7771
Takagi H, Abe A, Yoshida K, Kosugi S, Natsume S, Mitsuoka C, … Innan H (2013) QTL-seq: rapid mapping of quantitative trait loci in rice by whole genome resequencing of DNA from two bulked populations. Plant J 74(1):174–183
Tangahu BV, Sheikh Abdullah SR, Basri H, Idris M, Anuar N, Mukhlisin M (2011) A review on heavy metals (As, Pb, and Hg) uptake by plants through phytoremediation. Int J Chem Eng 2011
Tordoff GM, Baker AJM, Willis AJ (2000) Current approaches to the revegetation and reclamation of metalliferous mine wastes. Chemosphere 41(1):219–228
Truong PN, Foong YK, Guthrie M, Hung YT (2010) Phytoremediation of heavy metal contaminated soils and water using vetiver grass. In: Environmental bioengineering. Humana Press, Manila, Philippines. pp 233–275
Turpie J, Lannas K, Scovronick N, Louw A (2010) Wetland valuation volume I wetland ecosystem services and their valuation: a review of current understanding and practice. Water Research Commission, Republic of South Africa
Vaněk T, Podlipna R, Soudek P (2010) General factors influencing application of phytotechnology techniques. In: Application of phytotechnologies for

cleanup of industrial, agricultural, and wastewater contamination. Springer, Dordrecht, pp 1–13

Vara Prasad MN, de Oliveira Freitas HM (2003) Metal hyperaccumulation in plants: biodiversity prospecting for phytoremediation technology. Electron J Biotechnol 6(3):285–321

Vymazal J (2014) Constructed wetlands for treatment of industrial wastewaters: a review. Ecol Eng 73:724–751

Wang J, Chen C (2009) Biosorbents for heavy metals removal and their future. Biotechnol Adv 27(2):195–226

Wani SH, Sanghera GS, Athokpam H, Nongmaithem J, Nongthongbam R, Naorem BS, Athokpam HS (2012) Phytoremediation: curing soil problems with crops. Afr J Agric Res 7(28):3991–4002

Weick CW, Kaur S, Fernandez A, Jain RK (2003) Application of a method for selecting and evaluating environmental technologies with commercial potential. Int J Technol Transf Commer 2(4):399–428

Weyens N, Thijs S, Popek R, Witters N, Przybysz A, Espenshade J, … Gawronski SW (2015) The role of plant–microbe interactions and their exploitation for phytoremediation of air pollutants. Int J Mol Sci 16(10):25576–25604

Willey N (2006) Phytoremediation: methods and reviews, vol 23. Springer Science & Business Media, Dordrecht

Williams LE, Pittman JK, Hall JL (2000) Emerging mechanisms for heavy metal transport in plants. Biochim Biophys Acta Biomembr 1465(1):104–126

Wuana RA, Okieimen FE (2011) Heavy metals in contaminated soils: a review of sources, chemistry, risks and best available strategies for remediation. ISRN Ecol 2011:1–20

Xue K, van Nostrand JD, Vangronsveld J, Witters N, Janssen JO, Kumpiene J, … Zhou JZ (2015) Management with willow short rotation coppice increase the functional gene diversity and functional activity of a heavy metal polluted soil. Chemosphere 138:469–477

Yadav BK, Siebel MA, van Bruggen JJ (2011) Rhizofiltration of a heavy metal (lead) containing wastewater using the wetland plant Carex pendula. Clean Soil Air Water 39(5):467–474

Yin X, Yuan L, Liu Y, Lin Z (2012) Phytoremediation and biofortification: two sides of one coin. Springer, Dordrecht, pp 1–6

Yu T, Zhang Y, Meng W (2012) Environ Monit Assess 184: 4367–4832

Zaier H, Ghnaya T, Lakhdar A, Baioui R, Ghabriche R, Mnasri M, … Abdelly C (2010) Comparative study of Pb-phytoextraction potential in Sesuvium portulacastrum and Brassica juncea: tolerance and accumulation. J Hazard Mater 183(1):609–615

Zalesny RS, Wiese AH, Bauer EO, Riemenschneider DE (2006) Sapflow of hybrid poplar (Populus nigra L. × P. maximowiczii A. Henry 'NM6') during phytoremediation of landfill leachate. Biomass Bioenergy 30(8):784–793

Zhang C, Zhang P, Mo C, Yang W, Li Q, Pan L, Lee DK (2013) Cadmium uptake, chemical forms, subcellular distribution, and accumulation in Echinodorus osiris Rataj. Environ Sci Process Impact 15(7):1459–1465

Zhao FJ, McGrath SP (2009) Biofortification and phytoremediation. Curr Opin Plant Biol 12(3):373–380

Zhu YG, Pilon-Smits EA, Zhao FJ, Williams PN, Meharg AA (2009) Selenium in higher plants: understanding mechanisms for biofortification and phytoremediation. Trends Plant Sci 14(8):436–442

6 Concepts for Improving Phytoremediation by Plant Engineering

Contents

Abstract

Engineering plants with better metal tolerance and accumulation potential is imperative to emergent phytoremediators. Certain plants can hyperaccumulate metal ions that are lethal to nearly all organisms even at low dosages. This characteristic could be utilized for cleaning metal-polluted soils. Furthermore, the accretion of heavy metals by plants establishes both the micronutrient and the heavy metal concentration of our food chain. Intricate communications of transport and chelating processes manage the efficiency of metal uptake and storage. In current scenario, numerous fundamental steps have been recognized at the molecular level and facilitate us to commence transgenic advances to engineer the transition metal composition of plants. The utilization of genetic engineering to amend plants for metal uptake, transport, and sequestration may launch novel possibility for improving competence of phytoremediation. Preamble of genes governing chelation complexes and metal transporter can enhance metal uptake and sequestration. This results in transgenic plants with amplified detoxification and accumulation of heavy metals like cadmium, lead, mercury, arsenic, and selenium. An in-depth understanding pertaining to mechanisms of rhizosphere interaction, uptake, transport, and sequestration of metals in hyperaccumulator plants will guide us to

S. Jan, J.A. Parray, *Approaches to Heavy Metal Tolerance in Plants*,
DOI 10.1007/978-981-10-1693-6_6

designing new transgenic plants with enhanced remediation characteristics. As we will discover more genes associated with metal metabolism, assisted through the genome sequencing ventures, novel panorama will be announced for advancement of competent transgenic plant lines for phytoremediation.

Keywords

Metal chelator • Metal transporter • Metallothionein • Phytochelatin gene • Transgenic plant lines • Genetic engineering

6.1 Introduction

Numerous genes are implicated in metal uptake, translocation, and sequestration, and manipulation of several genes into contender plants is a potential approach for genetic manipulation of plants for enhanced phytoremediation characteristics (Clemens et al. 2002; Eapen and D'souza 2005). In context with advances, transgenic plants can be further advanced to amplify accrue elevated levels of metals aboveground parts of plants. Manipulation including overexpression of genes will result in improved metal uptake, translocation, sequestration, or intracellular targeting (Doty 2008). Genetic engineering of plants for production of metal chelators will advance the competence of plant for metal uptake (Tong et al. 2004). Conventional genetic investigations have established that only specific genes are accountable for metal tolerance (Kotrba et al. 2009). Increased levels of toxic metals, for instance, Cd, Pb, Cu, As, or Ni, in soils create a main ecological and human predicament. Phytoremediation is known to be a dynamic and eco-friendly technology to eliminate lethal metals from soils (Fulekar et al. 2009). The model phytoremediator should be rapidly emergent with increased biomass, be tolerant to and accrue increased content of lethal metals in the shoot, and be simple for cultivation and harvest. Even though natural hyperaccumulators can tolerate and accumulate high concentrations of lethal metals, they typically have lesser biomass, breed gradually, and are very difficult to cultivate (Yadav et al. 2010). Currently, with the progression of molecular biology, scientists can exploit genetic engineering to improve the metal accumulation capacity of rapid growth and high-biomass plants (Dessaux et al. 2016; Yu et al. 2016). Improvement of the metal accretion traits of plants via genetic engineering necessitates comprehensive perceptive of the biological processes implicated in metal acquisition by roots from soil sap, its transferred to shoots, and tolerance and accretion of elevated concentration of metals (Singh et al. 2016). However, the mechanisms of tolerance specifically in innate metal hyperaccumulators have improved metal acquirement, translocation, and tolerance and accretion capacity and thus are mainly important resource of prospective phytoremediation genes (Jagtap and Bapat 2015). It has been investigated that in order to advance the complex association of metal acquisition, genetic manipulation to amplify metal tolerance of phytoremediators seems to be exclusive alternative. Since higher capacity for metal accretion is basic requirement for phytoremediation, Ibañez et al. (2015) has reviewed the various research studies that have provided elucidative mechanism underlying metal tolerance, and several genes that impart tolerance against heavy metal have been identified. Research on microbes can offer an expensive model for comprehending and manipulating metal tolerance in plants since genes for metal acquisition are more advanced and varied than in plants (Benjamin et al. 2015). Most of genes exploited for increased metal tolerance have been originated from microbes except mammalian metallothionein and phytochelatin synthase. The method underlying the metal tolerance can be classified into two essential approaches. The most significant strategy is to regulate the heavy metal influx into cytoplasm by avoiding the metal translocation across plasma membrane via increased metal binding to the cell wall or by decreased uptake through modulated ion channels or via metal exclusion from cell using active efflux pumps, a method that is widespread in metal-tolerant bacteria (Tomar et al. 2015). Another strategy involves heavy metal detoxification through inactivation by chelation or transforming more toxic ion into relatively lower toxic ion or one that can be easily compartmentalized (Chen et al. 2013a). Despite the existence of varied metal tolerance pathways, the research on transgenic plants exhibiting increased metal accretion and tolerance capacity possesses amplified gene expression involved in biosynthetic mechanism of metal-binding proteins and peptides/genes that could transform more toxic ions to lesser toxic or just compartmentalize (Zhang et al. 2013). The perception of phytoremediation has recommended substantial significance in plant metal accumulation. Hyperaccumulation is main strategy for remediating metal-polluted water and soil resources as recommended by diverse elucidative research investigations which approve the exploitation of several plant and can modulate heavy metal toxicity as well as hyperaccumulate them in their aboveground parts (Dhankher et al. 2012). Plants best for phytoremediation should create numerous characteristics. These plants should exhibit intensive growth, increased plant productivity, and extensive root system harvestable and should tolerate and accrue a variety of heavy metals in their aboveground and yield outputs (Georgiev et al. 2012; Bell et al. 2014). However, single plant cannot exhibit all the features requisite for phytoremediation. Though, plant with extensive growth habit can be genetically tailored to attain status of hyperaccumulators. Modern advancement in shaping the molecular mechanism for metal accretion and tolerance by hyperaccumulators has been imperative and provides us with an intense scientific foundation to delineate some approach for attaining this objective (Peng et al. 2014). Molecular perceptive of plant metal accretion variables has several supplementary biotechnological inferences. In developing countries, deficiencies in trace metals are widespread and cause numerous human diseases in developing countries (Macek et al. 2012). Enhancement in zinc content can be induced through engineering Zn fortification in harvestable parts. In contrast, the common toxic nonessential element Cd is ingested by humans via plant-based products like tobacco smoke (Renault et al. 2014). Malagoli et al. (2015) reported significant discrepancy in Cd accretion are among diverse plant varieties, and elucidative explanation of molecular mechanism established could lead to advancement of functional markers or of crops with decrease toxic metal content. Finally, because signal transduction cascades are conjectured to have employed metal homeostasis features, an improved perceptive of plant metal management might also lead to new impending into other basic characteristics of plant physiology. In this chapter, we are relating significant characteristic of our existing perceptive of plant metal accumulation determinants and preferred engineering advances to improve metal accumulation competence (Pilon-Smits 2013). The major focal point is on Zn and Ni as the chief hyperaccumulated metals, on Cu as a principally toxic micronutrient, and on Cd, which is utilized as a "model" dispensable metal of important ecological significance in numerous studies (Rylott et al. 2015).

6.2 Classic Genetic Studies and Modern Approach for Improving Phytoremediation

Genetic analysis has offered indispensable perceptive of the functions implicated in phytoremediation which are relatively tiresome to infer at metabolic and biochemical levels. Genetic analysis proved a significant support for the recognition of genes for assembly of genetically modified plants for phytoremediation via traits of explicit plant varieties (Yang et al. 2015). To comprehend the source of tolerance and accretion of heavy metals and metalloids, conventional genetic research has been conquered for the recognition and elucidation of the genomic sequence in hyperaccumulators throughout genetic mapping and quantitative trait loci. This work was instigated and plasticized in two hyperaccumulators of model plants *Arabidopsis halleri* and *Thlaspi caerulescens* that were cross with non-hyperaccumulator ecotypes or congeners for studies of trait segregation (Seregin et al. 2014). This study established that trait overriding metal tolerance of each heavy metal is autonomous from accumulation of same metal, and tolerance is a polygenic trait. Courbot et al. (2007) established via bottom-up advance that gene HMA4 co-segregates with quantitative trait loci (QTL) while hauling out interspecific crosses between *Arabidopsis halleri* and *Arabidopsis lyrata* wherein tolerance and hyperaccumulation isolate as self-regulating traits while Cd tolerance co-segregates with Zn tolerance. Research group led by Martin Broadley had given twofold disparity in Cs concentration via QTL analysis in *Arabidopsis thaliana* (Lisec et al. 2008). QTL and mapping of entrant genes had assisted in isolation of progenies leading to mapping of putative QTLs on numerous chromosomes. Although conventional genetic studies are more or less complicated but consequential crosses always enhanced, the actual predicament is construing the segregation of the traits. Three QTLs and one QTL clearing up the overall discrepancy of 23.8–60.4 % were mapped for Zn and Cd concentration in shoots, correspondingly (Deniau et al. 2006). Soric et al. (2012) investigated disparity for Cu, Fe, K, Mn, Mg, and Sr concentrations in leaves of maize mapping population and determined the QTL linked with the metal concentrations. They established eleven important QTLs for six metal concentrations; of them, QTLs for Cu, Fe, and Mg were localized on chromosome 5 in the region of *ys1* gene underneath the participation of *ys1* in accumulation of these metals. The segregation of the QTL regions contributing Zn tolerance (*Zntol1*, *Zntol2*, *Zntol 3*) holds an enormous guarantee for enhancing the recognition of the major gene(s) accountable for the tolerance.

Recently, recognition of functional genes implicated in metal tolerance and metal homeostasis has been made promising through transcriptomics. Although an abundant data has been already generated by QTL studies during the last 5 years, transcriptomics provides accurate explanation of numerous genes offering chief function in metal tolerance and offers hint on the epistatic associations that exist among those genes. Until now transcript profiling has recognized further functional genes than proteomics owing its high-resolution competence and obvious information construal of thousands of genes at the same time. Proteomics has paved a way forward to recognition of only some proteins based on extractability and comparatively high expression level, and it perceives transcripts at exceptionally low levels (Chuh and Pratt 2015). However, proteomics can offer precise objectives whose function has been positively recognized in time. Consequently, proteins can be utilized in refining the genetic engineering advance by restraining the integer of objectives, defining accurately the goals themselves.

The main restrictions with genetic engineering comprise the efficiency of isolated gene and its expression in its innovative perspective. However, to progress the efficacy of expression, we must regard numerous features incorporated in so-called expression cassette. Accurate translation of coding sequence has to be accomplished which may necessitate alteration of codons. Furthermore, the gene expression of promoter sequence has to be competent in host species, possibly conferring tissue-specific expression, and other genes have to be modified consistent with host species. Genetic engineering in plants has an inadequate variety of pyramidization and has yet to be utilized to transfer complex traits, i.e., QTLs (Salvi and Tuberosa 2015).

6.3 Improved Metal Sequestration, Metal Transporters, and Allied Biomolecules via Genetic Engineering

Plants thriving in metal-polluted milieu might accrue lethal heavy metals and competently sort out these into diverse harvestable parts. Information collected from previous research articles designate that the separation of heavy metals at the entire plant stage could be generally divided into three classes (Ovečka and Takáč 2014). Anjum et al. (2015) illustrated alteration in plant metabolism to modulate plant resistance through tailoring metal translocation and their respective speciation. Kushwaha et al. (2015) established binary mechanism for tolerance against heavy metal toxicity that includes as follows:

(i) *Avoidance*, connecting diverse courses for averting lethal ions to accomplish their mark positions (Mehes-Smith et al. 2013).
(ii) *Tolerance*, against heavy metals within symplasm via complexation (Meharg 2005). The important metal complexation practices involve the production of phytochelatins and metal-chelating peptides (Yadav 2010). Diverse kinds of metal-chelating proteins and mechanism are involved in the intracellular ionic homeostasis (Valls and De Lorenzo 2002). The biosynthetic pathway of phytochelatins is mediated by enzyme, i.e., necessitates a protein stimulated by target metal (Mendoza-Cózatl et al. 2008). The heavy metal deposition prototype has been related with deforestation and the phytochelatin incidence (Nagy et al. 2011). It is also illustrated that some plants have a role as hyperaccumulators of explicit heavy metals due to their competent metal complexation ability (Boyd 2007). The recognition of metal complex associated with enzyme and along with illustrative products from the enzyme-mediated pathways is an emerging intriguing subject in ecological speciation investigation.

6.4 Genetic Manipulation of Metal-Sequestration Proteins and Peptides

Phytochelatins are ephemeral metal-stimulated sulfhydryl-rich peptides containing the broad configuration: (g-GluCys) n-Gly with n=2–11 (Cruz et al. 2005). Plant and fungi rendered to Ag, Bi, Cd, Cu, Hg, Ni, Sn, Sb, Te, W, Z, and anions, e.g., SeO_4^{-2}, SeO_3^{-2}, and AsO_4^{-3}. The phytochelatins are not stimulated by Na, Mg, Al, Ca, V, Cr, Mn, Fe, Co, and Cs (Cruz et al. 2005; Lima et al. 2006). Heavy metals form intracellular chelation complexes via coordinated thiol peptides having cysteine residues which lead to detoxification through vacuole sequestration (Yadav 2010). These metal detoxification complexes exhibit vivid nomenclature like cadystin, poly (g-glutamyl-cysteinyl) glycine, phytometallothioneins, g-glutamyl metal-binding peptide, g-glutamyl cysteinyl isopeptides, metallothiopeptide, metallopeptides, des glycyl peptides (g-glu-cys), and phytochelatins (Prasad 2012; Inouhe et al. 2015). Glutathione along with allied isoforms, i.e., homoglutathione and hydroxymethyl-glutathione, are widespread thiols found across plant kingdom (Pivato et al. 2014). Seth et al. (2012) reported that sulfur-containing complexes ameliorate heavy metal toxicity possibly by modulating glutathione metabolism (Seth et al. 2012). Metal-stimulated phytochelatin production decline glutathione levels within cell. Maughan et al. (2010) confirmed alteration in thiol pools of heavy metal-contaminated plants can be synchronized through intermediates and effectors of glutathione synthesis. Hence, there exists an alternative of elevating stage of metal chelates in plants through accretion of GSH levels. This is accomplished by overexpression of enzymes that govern biosynthesis of GSH. Transgenic poplar has been developed by introducing amplified gene expression of bacterial enzyme γ-glutamylcysteine synthetase which regulates GSH biosynthesis (Yadav et al. 2010). The transgenic plants reveal increased GSH production and being investigated for their heavy metal uptake and tolerance. Cruz et al. (2005) swapped strategy by investigating metal-binding affinity of synthetic phytochelatin analogs $(Glu\text{-}Cys)_nGly$ peptides. These synthetic peptides have been revealed to display metal-chelating characteristics comparable to natural phytochelatins (Prévéral et al. 2009). Synthetic genes coding for these deviant peptides are being investigated currently intending the generation of metal-resistant transgenic plants (Dhankher et al. 2002). In an illustrative review where class I and

II metallothioneins were recognised as supplementary assortment of non-enzymatic polypeptides having low molecular mass (6–7 kDa) creamed with cysteinyl residues exhibiting high complexation competence. These protein complexes exhibit high tolerance to thermocoagulation and acid precipitation due to nonexistence of disulfide bonding (Zhou et al. 2005). Even though MTs are widespread in the animals, numerous plants are investigated to include MT genes, viz., pea (*Pisum sativum*), soybean (*Glycine max*), thale cress (*Arabidopsis thaliana*), monkey flower (*Mimulus guttatus*), maize (*Zea mays*), barley (*Avena sativa*), wheat (*Triticum aestivum*), castor bean (*Ricinus communis*), and rapeseed (*Brassica napus*) (Cobbett and Goldsbrough 2002). Cloning of metallothionein genes and their integration into an appropriate plant species leads to high metal tolerance. Tobacco and oil seeds with better Cd tolerance were developed via transfer of human MT-2 gene, while MT gene transfer from pea resulted in Cu tolerant in tobacco and *A. thaliana* (Kumar et al. 2012). Expression of metallothionein (MT) gene and high transcription of genes for glutathione synthesis have been related with Cu and Cd tolerance in *Arabidopsis thaliana* (Lin et al. 2013). Constitutive expression of diverse MT genes like MTI, human MTIA, human MTII, yeast CUPI, pea *Ps*MT A, and *TaMT3* when transferred to *Nicotiana tabacum*, *Brassica* species, and *Arabidopsis thaliana* led to increased cadmium (Cd) tolerance (Zhou et al. 2014). Expression of metallothionein gene and increased transcription of genes of glutathione synthesis have been correlated with Cu and Cd tolerance in *Arabidopsis thaliana* (Yan et al. 2015). *Arabidopsis thaliana* exhibited improved Cu and Al tolerance when gene encoding glutathione S-transferase was transferred from tobacco (Ezaki et al. 2000). Current studies also exhibited that transgenic plants overexpressing BcMT1 *and* BcMT2 gene from *Brassica campestris* in *Arabidopsis* enhance Cd and Cu stress tolerance by increasing heavy metal concentrations and lower ROS accumulation (Lv et al. 2013). Similarly, Chaturvedi et al. (2014) demonstrated that transgenic tobacco plants overexpressing SbMT-2 gene improve metals (Zn^{2+}, Cu^{2+}, and Cd^{2+}) and abiotic stress (salt and osmotic) tolerance through high expression of functional antioxidant enzyme encoding genes, SOD, POD, and APX and by altering ROS detoxification. Improved generation of Fe chelator, deoxymugeinic acid (DMA) in rice, occurred through the transfer of NicotinamideN-methyl transferase (NNMT) gene. Numerous researches have demonstrated the improvement of metallothionein genes and those genes encoding for the enzymes essential in sulfur metabolism and histidine biosynthesis, 1-aminocyclopropane-1-carboxylate (ACC) deaminase, phytochelatin synthase, Hg^{2+}reductase, arsenate reductase, and aldolase/aldehyde reductase (Iqbal et al. 2014).

6.5 Genetic Engineering for Encoding Metal Ion Transporters

The development of metal/metalloid uptake and its transport mechanism is complicated and diverse (Foucault et al. 2013). The transport of metal and its screening to plant organs particularly its accretion into vacuoles require metal transporters. The competent method which could result in improved metal tolerance will be achieved via genetic manipulation of metal transporters (Yang et al. 2015). The discriminatory pathways that necessitate to be focused in view of their relevance to phytoremediation would include exploitation of metal transporter at three levels: (a) metal uptake from soil to root, (b) root-to-shoot metal translocation, and (c) metal sequestration into vacuoles or cell wall. Despite an ample research on genes encoding transporters across plasma membrane of root epidermal cells, none of research in genetic engineering had been focused on the enhanced absorption of toxic metals from soil solution to roots (Xie et al. 2015; Jan et al. 2016). Metal transport from the cytosol to the vacuole is considered to be an imperative constituent of ion tolerance and its prospective for its utilization in phytoremediation:

(i) *CPx-type ATPases*: They include IB class of heavy metal transporters since they contribute to the widespread characteristic of a preserved intramembranous cysteine-protein-cysteine, cysteine-proline-histidine, or cysteine-proline-serine motif utilized in heavy metal transduction (Uversky 2013). CPx-type heavy metal ATPases utilize ATP to pump a diverse of charged substances across cell membrane and have an imperative purpose in the transport of necessary in addition to toxic metals, for instance, Cu, Zn, Cd, and Pb, across the cell membrane (Guo et al. 2015). CPx-type ATPases may present in the plasma membrane and purpose as efflux pumps, eliminate potentially toxic metals from cytoplasm, or may also present at diverse intracellular membranes and accountable for compartmentalization of heavy metals. Heavy metal transporters comprise CPx-type or ATPase, implicated in the entire metal ion homeostasis and tolerance found to be accountable for compartmentalization of heavy metals in plants (Hassan and Aarts 2011). The ectopic overexpression of AtHMA4 which is an *Arabidopsis thaliana* P-1 BATPase enhanced root growth in the presence of Zn, Cd, and Co (Morel et al. 2009). Amplified expression of *HMA4* in *A. halleri* or *T. caerulescens* and in tobacco exhibited improved Cd and Zn tolerance and hyperaccumulation phenotype of these species (Grispen et al. 2011).

(ii) *Natural resistance-associated macrophage protein (Nramp)*: NRAMPs are proton-driven metal ionoots and shoots. These transporters transfer widespread variety of metals, for example, Mn, Zn, Cu, Fe, Cd, Ni, and Co across membranes (Nevo and Nelson 2006). In *Arabidopsis*, heterologous expression of *AtNRAMP1*, *AtNRAMP3*, and *AtNRAMP4* in yeast mutants indicated that these proteins could transport Fe, Mn, and cadmium (Cellier et al. 2012; Ehrnstorfer et al. 2014). *AtNRAMP3* and *AtNRAMP4* are accountable for mobilization of vacuolar Fe reserves. Rice OsNRAMP1 rescues the growth of a Fe-defective yeast mutant and is associated to Cd accumulation in rice (Takahashi et al. 2012). Currently, description of rice OsNRAMP5 discovered its contribution in Mn, Fe, and Cd uptake and transport (Ishimaru et al. 2012). Several SMF, which located for suppressor of mitochondria import purpose and localized in the yeast mitochondrial membrane. SMF1, SMF2, and SMF3 are yeast homologs of the Nramp proteins with 51–54 % resemblance in amino acid sequence and 33–36 % resemblance to DCT1 and implicated in uptake of Mn and Zn into the cytoplasm (Tamayo et al. 2014; Khan et al. 2015).

(iii) *Cation diffusion facilitator (CDF)*: It consists of protein family implicated in cation efflux away from cytoplasmic compartment either transversely or into cellular compartments, for instance, vacuoles (Chen et al. 2013b). The CzcD from *Ralstonia eutropha* mediates Zn^{2+}efflux, and the proteins COT1 and ZRC1 from *S. cerevisiae* when overexpressed confer Co and Zn/Cd tolerance in these plants, respectively (Zhao et al. 2012). This recommends the function of these proteins in metal sequestration.

(iv) *Zinc-iron permease (ZIP)*: This protein family includes metal ion transporters that transport metals, for instance, Fe, Zn, and Cd, having 15 members in *Arabidopsis* genome. Zinc accretion in roots of *Arabidopsis thaliana* was improved by two-fold via transfer of Zn transporter *ZAT* gene from *Thlaspi goesingense* (Ahmad 2015). Improved accumulation of Ca, Cd, and Mn in tobacco was stimulated via the transfer of calcium vacuolar transporter *CAX-2* from A. *thaliana* (Sarangi et al. 2010). Correspondingly, the transfer of another transporter gene *NtCBP4* which codes for calmodulin-binding protein improved nickel tolerance (Baligarx 2012). The ZIP transporters were illustrated in *Arabidopsis* soybean and rice (Milner et al. 2013). Description and recognition of cDNA in *Medicago truncatula* capitulated six new metal transporters (*Mt ZIP*) exhibiting improved resemblance with ZIP family

(Adams et al. 2012). *MtZIP3* and *MtZIP4* were downregulated in Mn inadequate media, while *MtZIP5* expression was upregulated in leaves (King 2011). The IRT1 transporter segregated from *Arabidopsis* displays an extensive substrate variety assisting in transport of divalent ions from soil like Mn, Zn, Fe(II), and probably Cd^{2+} (Barberon et al. 2011).

Cell walls and vacuoles contribute significant function in detoxifying metals particularly in Ni-, Zn-, or Cd-hyperaccumulating plant species (Rascio and Navari-Izzo 2011). Metal accretion in the vacuole may reduce switchover border in metal and metabolic process. *Arabidopsis thaliana* acquire nine high-affinity phosphate transporters (PHT) which exhibit varied reaction toward arsenate signifying their intervention with phosphate signalling (AsV) (LeBlanc et al. 2013). The mutation in one of the nodulin intrinsic protein LSi1 which is a silicon transporter leads to 60 % loss in As III influx in rice roots (Ma and Yamaji 2015). An additional NIP family member LSi2 display maximum As III loading inclination into xylem, and approximately 50 % decrease in arsenic accretion of shoots has occurred due to T-DNA in LSi2 locus (Liu et al. 2014). Mutant lines in *Arabidopsis* population have been exploited broadly for segregation of genes for Pb and Mn tolerance leading to recognition of Pb-tolerant mutants. The Fe content of tobacco plants was enhanced by 1.5-fold when FRE1 and FRE2 gene encoding ferric reductase were transferred from yeast (Xu et al. 2014). Recently, wide-ranging investigation has been established on bioengineering of plants tolerant to declining concentrations of toxic metals, for instance, Se, Cd, As, Hg, and Pb.

6.6 Genetic Engineering of Enzymes to Enhance Phytovolatilization

Mercury bioremediation in wetlands has been managed by production of an extensive rooting system in yellow poplar and cottonwood through phytovolatilization. The progression in metabolic engineering leads to the liberation of transgenic plants having effectual of both MerB (organomercurial lyase) and MerA (mercuric reductase) enzymes for detoxifying harmful mercury compounds (i.e., ionic mercury Hg(II), methylmercury, and phenylmercuric acetate PMA) (Jagtap and Bapat 2015). Elevated intensity of tolerance was demonstrated in transgenic tobacco plants that accrue more Hg in the shoot when grown in medium affluent in phenylmercuric acetate (PMA) or $HgCl_2$ (Pinto et al. 2014). Though the detoxification of mercury has been effectively accomplished through phytovolatilization, still it cannot validate exit of volatile mercury from complete globe. Consequently, the need for more compact strategy was stressed that could initiate hyperaccumulation of Hg(II) in aboveground parts which could be later harvested to avert liberation of Hg(0) in atmosphere. These plants were extra tolerant to Hg and can accrue elevated absorption of methylmercury than plants that communicate single transgene.

Transgenic *Brassica juncea* displayed improved selenate decrease and improvement of enzymes ATP sulfurylase (APS), implicated in selenate-to-selenite conversion (Pilon-Smits et al. 2014). These transgenic plants demonstrate two- to threefold higher Se volatilization rates than wild type and 1.5-fold extra sulfur. This characteristic was extended in Indian mustard by overexpression of gene that codes for primary enzyme in the exchange of SeCys to SeMet specifically cystathionine gamma synthase. These transgenics accrue 40 % less Se than wild types due to their improved volatilization. These transgenic lines were enhanced by hubbing on the anticipation of SeCys insertion into proteins. Expression of selenocysteine lyase (SL) was stimulated in *Arabidopsis thaliana* and Indian mustard via homolog called cpNifS (Pilon-Smits and Elizabeth 2015). This enzyme breaks SeCys into alanine and Se. The consequential transgenics exhibited declined Se insertion into proteins, amplified Se tolerance, and twofold increased in Se accumulation evaluated to non-transformed plants. Further, approach was extended through overexpression of enzyme SeCys methyltransferase in *Arabidopsis thaliana* or *Brassica juncea* which was obtained from hyperaccumulator *Arabidopsis bisulcatus*. The SeCysMT transgenics reveal augmented Se volatilization leading to increased tolerance and amended Se accumulation in the form of methyl-SeCys. *Brassica juncea* which acquire both ATP sulfurylase (APS) and SeCys methyltransferase (SMT) transgenic trait can productively accrue ninefold higher of Se evaluated to non-transgenic types. A significant transgenic that was developed for effective accumulation of selenium was tailored through the introduction of selenocysteine methyltransferase gene of *Astragalus bisulcatus* (two-grooved poison vetch) into *Arabidopsis thaliana* (thale cress) to improve the expression of Se-methylselenocysteine and γ-glutamylmethylselenocysteine in shoots (Pilon-Smits and LeDuc 2009). In addition, novel plant *Cardamine hupingshanesis* has been recognized as hyperaccumulator for selenium (Yuan et al. 2013; Wu et al. 2015).

6.7 Improving Zinc Phytoremediation Efficiency

For most effective phytoremediation, plants utilized should possess both huge biomass production and amplified metal accumulations in shoots (Vamerali et al. 2010). Though, no plant species meet such criteria. For the Zn remediation management, plants were added with phytohormones to advance root and shoot development for augmented biomass productivity. By means of biodegradable chelating agents, it can advance the removal of heavy metals from soil. However, using chelating agents for enhancing phytoremediation should be done with resistance to evade their intervention with microbial diversity, heavy metal leaching, and accretion of intricate organic chelating agents in the environment. Genetic engineering is hitherto to investigate the alternative for enhancement of plants that could be exploited for Zn hyperaccumulation. Currently, scientists are advancing toward the recognition of genes implicated in regulation of metal transport system transversely in plasma membrane, improved production of natural chelators like metallothioneins (MTs) and phytochelatins (PCs), and alteration of the Zn metabolic pathways.

Comprehensive research is essential for improved molecular and genetic mechanism in Zn hyperaccumulators required for ecological and effective phytoremediation. Most of Zn-tolerant lines were acquired for enhancing the micronutrient composition in food or staple crops, for instance, rice as biofortification (Bashir et al. 2013; Tan et al. 2015). Traditional breeding can enhance Zn tolerance and its mobility across phloem of fruits, seeds, and tuber crops. Furthermore, traditional breeding and genetic engineering have targeted on the recognition of genes implicated in execution and regulation of physiological activities, for instance, Zn uptake, translocation, distribution, and sequestration in plants predominantly in preferred Zn-tolerant and Zn hyperaccumulators. Studies on adaptation to Zn deficiency in *Arabidopsis thaliana* discovered the possible function of transcription factors in regulation of Zn homeostasis in plant (Claus et al. 2013). The overexpression of bZIP19 and bZIP23 transcription factors could advance Zn accumulation in harvestable plant parts by inducing constitutive expression of a group of Zn-deficiency responses (Assuncao et al. 2010). Improved contents of Zn in leaves of *Arabidopsis thaliana* were examined when the expression of *AtHMA2* that activates Zn^{2+} efflux across the membranes of root cells is reduced or overexpressing the gene encoding *AtHMA4*, which is thought to load Zn into the xylem (Takahashi et al. 2011). Overexpression of *HvNAS1* in tobacco augmented Zn content in leaf and seed enhanced from 16 to 39 mg kg^{-1} DW and from 20 to 35 mg kg^{-1}DW, correspondingly. Several crops accrued Zn contents at increased rates like cassava with roots Zn of about 40 mg kg^{-1} DW (Sayre et al. 2011), the brown rice with 56–95 mg Zn kg^{-1} DW (Johnson et al. 2011), and barley grain with 85 mg Zn kg^{-1} DW have been demonstrated. Devastation in root absorption of Zn, decreasing solubility, annual rainfall, and high pH can considerably decrease plant Zn concentrations. Zn deficiency in soils of countries like China, India, Iran, Pakistan, and Australia hampers the addition of Zn in edible parts (Jan et al. 2016). Thus, genetic strategies should be expanded for efficient sequestration of Zn leading to development of extra practical biofortification. Consequently, an appropriate blend of breeding and fertilizer method is an exceptional contrary strategy to advance Zn deficiency in soil. Zn phytoextraction can be analyzed extra in cereal plants with an elevated biomass and adequate contents of Zn in edible parts; this could improve the trace element accessibility in rhizosphere and its translocation toward aerial and grain tissues.

Recognition and functional description of new proteins implicated in vacuolar sequestration of heavy metals and cross-membrane transport had also accelerated the continual exploitation of plants for enhanced phytoremediation purposes. Comprehensively, investigations of molecular mechanisms of pollutant uptake, accretion, detoxification, and exclusion reveal that some of these mechanisms are species specific, and whole perceptive of plant tolerance to pollution are still unraveled. An improved perceptive of the molecular base of the pathways and genes implicated in the degradation of pollutants is consequently vital. In addition, analysis and discovery of new genes appropriate for phytoremediation are indispensable, and accumulators require to be engineered with several stacked genes so as to generate elevated accumulator plants that can be exploited for more efficient phytoremediation of contaminated sites. Genetic engineering should target further on vacuolar sequestration, which is an imperative module of metal hypertolerance and putative hyperaccumulation. Exertion should be made for advancing root uptake, xylem loading, vacuolar transport, and mobilization of heavy metals. Comprehensive perceptive of molecular mechanism indispensable for hyperaccumulation and biochemical categorization of root exudates needs to be explicated. Several other plant species require to be investigated for their tolerance and accumulation. Genetic engineering for enhancing the metal hyperaccumulators by altering oxidation states, escalating the metal-sequestration proteins and recognition of transport proteins, for instance, ZIP and ZAT, could lead to improved and extra-efficient phytoremediation technology. Fortification and perpetuation of hyperaccumulators which are frequently established in geographically remote areas and exposed to wreckage via mining upheaval require an imperative consideration.

References

Adams JP, Adeli A, Hsu CY, Harkess RL, Page GP, Pamphilis CWD, Yüceer C (2012) Characterization of Poplar ZIP family members ZIP1.2 and ZNT1. J Plant Gene Transgen 3(1):1–13

Ahmad P (2015) Plant metal interaction: emerging remediation techniques. Elsevier, Amsterdam

Anjum NA, Hasanuzzaman M, Hossain MA, Thangavel P, Roychoudhury A, Gill SS, … Duarte AC (2015) Jacks of metal/metalloid chelation trade in plants—an overview. Front Plant Sci 6:192. doi: 10.3389/fpls.2015.00192

Assuncao AGL, Herrero E, Lin YF (2010) *Arabidopsis thaliana* transcription factors bZIP19 and bZIP23 regulate the adaptation to zinc deficiency. Proc Natl Acad Sci U S A 107:10296–10301

Baligarx VC (2012) Mechanisms of nickel uptake and hyperaccumulation by plants and implications for soil remediation. Adv Agron 117:117

Barberon M, Zelazny E, Robert S, Conejero G, Curie C, Frim J, Vert G (2011) Monoubiquitin-dependent endocytosis of the iron-regulated transporter 1 (IRT1) transporter controls iron uptake in plants. Proc Nat Acad Sci USA 108(32):E450–E458

Bashir K, Takahashi R, Hiromi N, Nishizawa NK (2013) The road to micronutrient biofortification of rice: progress and prospects. Fron Plant Sci 4:15

Bell TH, Joly S, Pitre FE, Yergeau E (2014) Increasing phytoremediation efficiency and reliability using novel omics approaches. TRENDS in Biotechnology 32(5):271–280

Benjamin SR, de Lima F, Rathoure AK (2015) Genetically engineered microorganisms for bioremediation processes: GEMs for bioremediaton. In: Toxicity and waste management using bioremediation, vol 113. Information Resources Management Association, Hershey

Boyd RS (2007) The defense hypothesis of elemental hyperaccumulation: status, challenges and new directions. Plant and Soil 293(1–2):153–176

Cellier A, Francou C, Houot S, Ballini C, Gauquelin T, Baldy V (2012) Use of urban composts for the regeneration of a burnt Mediterranean soil: a laboratory approach. J Environ Man 95:S238–S244

Chaturvedi AK, Patel MK, Mishra A, Tiwari V, Jha B (2014) The SbMT-2 gene from a halophyte confers abiotic stress tolerance and modulates ROS scavenging in transgenic tobacco. PLoS One 9(10):e111379. doi:10.1371/journal.pone.0111379

Chen Y, Xu W, Shen H, Yan H, Xu W, He Z, Ma M (2013a) Engineering arsenic tolerance and hyperaccumulation in plants for phytoremediation by a PvACR3 transgenic approach. Environ Sci Technol 47(16):9355–9362

Chen Z, Fujii Y, Yamaji N, Masuda S, Takemoto Y, Kamiya T, Ueno D (2013b) Mn tolerance in rice is mediated by MTP8. 1, a member of the cation diffusion facilitator family. J Exper Bot 64(14):4375–4387

Chuh KN, Pratt MR (2015) Chemical methods for the proteome-wide identification of post translationally modified proteins. Cur Opinion in Chem Bio 24:27–37

Claus J, Bohmann A, Chavarria-Krauser A (2013) Zinc uptake and radial transport in roots of *Arabidopsis thaliana*: a modelling approach to understand accumulation. Ann Bot 112:369–380

Clemens S, Palmgren MG, Krämer U (2002) A long way ahead: understanding and engineering plant metal accumulation. Trends Plant Sci 7(7):309–315

Cobbett C, Goldsbrough P (2002) Phytochelatins and metallothioneins: roles in heavy metal detoxification and homeostasis. Annu Rev Plant Biol 53(1):159–182

Courbot M, Willems G, Motte P, Arvidsson S, Roosens N (2007) A major quantitative trait locus for cadmium tolerance in Arabidopsis halleri colocalizes with HMA4, a gene encoding a heavy metal ATPase. Plant Phys 144:1052–1065

Cruz BH, Díaz-Cruz JM, Ariño C, Esteban M (2005) Complexation of heavy metals by phytochelatins: voltammetric study of the binding of Cd2+ and Zn2+ ions by the phytochelatin (γ-Glu-Cys) 3Gly assisted by multivariate curve resolution. Environ Sci Technol 39(3):778–786

Deniau AX, Pieper B, Ten Bookum WM, Lindhout P, Aarts MGM, Schat H (2006) QTL analysis of cadmium and zinc accumulation in the heavy metal hyperaccumulator *Thlaspi caerulescens*. Theor Appl Gen 113(5):907–920

Dessaux Y, Grandclément C, Faure D (2016) Engineering the Rhizosphere. Trends Plant Sci 21:266–278

Dhankher OP, Li YJ, Rosen BP, Shi J, Salt D, Senecoff JF, Sashti NA, Meagher RB (2002) Engineering tolerance and hyperaccumulation of arsenic in plants by combining arsenate reductase and gamma-glutamylcysteine synthetase expression. Nat Biotechnol 20:1140–1145

Dhankher OP, Pilon-Smits EA, Meagher RB, Doty S (2012) Biotechnological approaches for phytoremediation. In: Plant biotechnology and agriculture. Elsevier, Amsterdam, pp 309–328

Doty SL (2008) Enhancing phytoremediation through the use of transgenics and endophytes. New Phytologist 179(2):318–333

Du J, Yang JL, Li CH (2012) Advances in metallotionein studies in forest trees. Plant Omics 5(1):46

Eapen S, D'souza SF (2005) Prospects of genetic engineering of plants for phytoremediation of toxic metals. Biotechnol Adv 23(2):97–114

Ehrnstorfer IA, Geertsma ER, Pardon E, Steyaert J, Dutzler R (2014) Crystal structure of a SLC11 (NRAMP) transporter reveals the basis for transition-metal ion transport. Nat Struc Mole Bio 21(11):990–996

Ezaki B, Gardner RC, Ezaki Y, Matsumoto H (2000) Expression of aluminum-induced genes in transgenic Arabidopsis plants can ameliorate aluminum stress and/or oxidative stress. Plant Physiol 122(3):657–666

Foucault Y, Leveque T, Xiong T, Schreck E, Austruy A, Shahid M, Dumat C (2013) Green manure plants for remediation of soils polluted by metals and metalloids: ecotoxicity and human bioavailability assessment. Chemo 93(7):1430–1435

Fulekar MH, Singh A, Bhaduri AM (2009) Genetic engineering strategies for enhancing phytoremediation of heavy metals. Afr J Biotechnol 8(4):529–535

Georgiev MI, Agostini E, Ludwig-Müller J, Xu J (2012) Genetically transformed roots: from plant disease to biotechnological resource. Trends Biotechnol 30(10):528–537

Grispen VM, Hakvoort HW, Bliek T, Verkleij JA, Schat H (2011) Combined expression of the Arabidopsis metallothionein MT2b and the heavy metal transporting ATPase HMA4 enhances cadmium tolerance and the root to shoot translocation of cadmium and zinc in tobacco. Environ Exp Bot 72(1):71–76

Guo J, Green BR, Maldonado MT (2015) Sequence analysis and gene expression of potential components of copper transport and homeostasis in *Thalassiosira pseudonana*. Protist 166(1):58–77

Hassan Z, Aarts MG (2011) Opportunities and feasibilities for biotechnological improvement of Zn, Cd or Ni tolerance and accumulation in plants. Environmental and Experimental Botany 72(1):53–63

Ibañez SG, Paisio CE, Oller ALW, Talano MA, González PS, Medina MI, Agostini E (2015) Overview and new insights of genetically engineered plants for improving phytoremediation. In: Phytoremediation.

Management of environmental contaminants, vol 1. Springer International Publishing, pp 99–113. doi:10.1007/978-3-319-10395-2_8

Inouhe M, Sakuma Y, Chatterjee S, Datta S, Jagetiya BL, Voronina AV, … Gupta DK (2015) General roles of phytochelatins and other peptides in plant defense mechanisms against oxidative stress/primary and secondary damages induced by heavy metals. In: Reactive oxygen species and oxidative damage in plants under stress. Springer International Publishing, Cham, pp 219–245

Iqbal M, Ahmad A, Ansari MKA, Qureshi MI, Aref IM, Khan P.R, … Hakeem KR (2014) Improving the phytoextraction capacity of plants to scavenge metal (loid)-contaminated sites. Environ Rev 23(1):44–65

Ishimaru Y, Takahashi R, Bashir K, Shimo H, Senoura T, Sugimoto K, Nishizawa NK (2012) Characterizing the role of rice NRAMP5 in manganese, iron and cadmium transport. Sci Rep 2:286

Jagtap UB, Bapat VA (2015) Genetic engineering of plants for heavy metal removal from soil. In: Heavy metal contamination of soils. Springer International Publishing, Cham, pp 433–470

Jan S, Kamili AN, Parray JA, Bedi YS, Ahmad P (2016) Microclimatic variation in UV perception and related disparity in tropane and quinolizidine alkaloid composition of Atropa acuminata, Lupinus polyphyllus and Hyoscyamus niger. J Photochem Photobiol B Biol 161:230–235

Johnson AAT, Kyriacou B, Callahan DL (2011) Constitutive over expression of the OsNAS gene family reveals single-gene strategies for effective iron- and zinc-biofortification of rice endosperm. PLoS One 6(9):24476

Khan N, Ryu KY, Choi JY, Nho EY, Habte G, Choi H, Kim KS (2015) Determination of toxic heavy metals and speciation of arsenic in seaweeds from South Korea. Food Chem 169:464–470

King I (2011) The role of Zip Superfamily of metal transporters in chronic diseases, purification & characterization of a bacterial Zip Transporter: Zupt. Wayne State University theses. Paper 63

Kotrba P, Najmanova J, Macek T, Ruml T, Mackova M (2009) Genetically modified plants in phytoremediation of heavy metal and metalloid soil and sediment pollution. Biotechnol Adv 27(6):799–810

Kumar G, Kushwaha HR, Panjabi-Sabharwal V, Kumari S, Joshi R, Karan R, … Pareek A (2012) Clustered metallothionein genes are co-regulated in rice and ectopic expression of OsMT1e-P confers multiple abiotic stress tolerance in tobacco via ROS scavenging. BMC Plant Biol 12(1):107

Kushwaha A, Rani R, Kumar S, Gautam A (2015) Heavy metal detoxification and tolerance mechanisms in plants: implications for phytoremediation. Environmental Reviews 23(999):1–13

LeBlanc MS, McKinney EC, Meagher RB, Smith AP (2013) Hijacking membrane transporters for arsenic phytoextraction. J Biotech 163(1):1–9

Lima AIG, Pereira SIA, Figueira EMDAP, Caldeira GCN, de Matos Caldeira HDQ (2006) Cadmium detoxification in roots of Pisum sativum seedlings: relationship between toxicity levels, thiol pool alterations and growth. Environmental and experimental botany 55(1):149–162

Lin CY, Trinh NN, Fu SF, Hsiung YC, Chia LC, Lin CW, Huang HJ (2013) Comparison of early transcriptome responses to copper and cadmium in rice roots. Plant Mol Biol 81(4–5):507–522

Lisec J, Meyer RC, Steinfath M, Redestig H, Becher M, Witucka-Wall H, Willmitzer L (2008) Identification of metabolic and biomass QTL in *Arabidopsis thaliana* in a parallel analysis of RIL and IL populations. The Plant J 53(6):960–972

Liu J, Shang W, Zhang X, Zhu Y, Yu K (2014) Mn accumulation and tolerance in *Celosia argentea* Linn.: a new Mn-hyperaccumulating plant species. J Hazard Mat 267:136–141

Lv Y, Deng X, Quan L, Xia Y, Shen Z (2013) Metallothioneins BcMT1 and BcMT2 from *Brassica campestris* enhance tolerance to cadmium and copper and decrease production of reactive oxygen species in *Arabidopsis thaliana.* 367: 507–519

Ma JF, Yamaji N (2015) A cooperated system of silicon transport in plants. Trends Plant Sci 20(7):435–42

Macek T, Novakova M, Kotrba P, Viktorova J, Lovecká P, Fiser J, … Mackova M (2012) 23 genetically modified plants designed for phytoremediation of toxic organic and inorganic contaminants. In: Phytotechnologies: remediation of environmental contaminants. CRC Press, Boca Raton, pp 415.

Malagoli M, Schiavon M, dall'Acqua S, Pilon-Smits EA (2015) Effects of selenium biofortification on crop nutritional quality. Front Plant Sci 6:280

Maughan SC, Pasternak M, Cairns N, Kiddle G, Brach T, Jarvis R, … Salcedo-Sora E (2010) Plant homologs of the Plasmodium falciparum chloroquine-resistance transporter, PfCRT, are required for glutathione homeostasis and stress responses. Proc Nat Acad Sci 107(5):2331–2336

Meharg AA (2005) Mechanisms of plant resistance to metal and metalloid ions and potential biotechnological applications. In: Root physiology: from gene to function. Springer, Dordrecht, pp 163–174

Mehes-Smith M, Nkongolo K, Cholewa E (2013) Coping mechanisms of plants to metal contaminated soil. In: Environmental change and sustainability. InTech, ISBN, pp 978–953

Mendoza-Cózatl DG, Springer F, Torpey JW, Komives EA, Kehr J, Schroeder JI (2008) Identification of high levels of phytochelatins, glutathione and cadmium in the phloem sap of Brassica napus. A role for thiol-peptides in the long-distance transport of cadmium and the effect of cadmium on iron translocation. Plant J 54(2):249–259

Milner MJ, Seamon J, Craft E, Kochian LV (2013) Transport properties of members of the ZIP family in plants and their role in Zn and Mn homeostasis. J Exper Bot 64(1):369–381

Morel M, Crouzet J, Gravot A, Auroy P, Leonhardt N, Vavasseur A et al (2009) AtHMA3, a P1B-ATPase allowing Cd/Zn/Co/Pb vacuolar storage in Arabidopsis. Plant Physiol 149:894–904

Nagy KL, Manceau A, Gasper JD, Ryan JN, Aiken GR (2011) Metallothionein-like multinuclear clusters of mercury (II) and sulfur in peat. Environ Sci Technol 45(17):7298–7306

Nevo Y, Nelson N (2006) The NRAMP family of metal-ion transporters. Biochim et Biophy Acta (BBA)-Mol Cell Res 1763(7):609–620

Ovečka M, Takáč T (2014) Managing heavy metal toxicity stress in plants: biological and biotechnological tools. Biotechnol Adv 32(1):73–86

Öztürk M, Ashraf M, Aksoy A, Ahmad MSA, Hakeem KR (2016) Plants, pollutants and remediation. Springer, Dordrecht

Peng R, Fu X, Tian Y, Zhao W, Zhu B, Xu J, … Yao Q (2014) Metabolic engineering of Arabidopsis for remediation of different polycyclic aromatic hydrocarbons using a hybrid bacterial dioxygenase complex. Metab Eng 26:100–110

Pilon-Smits EAH (2013) Engineering plant selenium accumulation: potential uses and ecological impacts. In: Selenium in the environment and human health. CRC Press, Boca Raton, pp 217

Pilon-Smits EAH, Elizabeth A (2015) Selenium in plants. In: Progress in botany. Springer International Publishing, Berlin/New York, pp 93–107

Pilon-Smits EAH, LeDuc DL (2009) Phytoremediation of selenium using transgenic plants. Curr Opin Biotechnol 20:207–212

Pilon-Smits EAH, Banuelos GS, Parker DR (2014) Uptake, metabolism, and volatilization of selenium by terrestrial plants. In: Salinity and drainage in San Joaquin Valley, California. Springer, Dordrecht, 147–164

Pinto AP, de Varennes A, Fonseca R, Teixeira DM (2014) Phytoremediation of soils contaminated with heavy metals. Phytoremediation: Management of Environmental Contaminants 1:133

Pivato M, Fabrega-Prats M, Masi A (2014) Low-molecular-weight thiols in plants: functional and analytical implications. Arch Biochem Biophys 560:83–99

Prasad AS (2012) Discovery of human zinc deficiency: 50 years later. J Trace Elem Med Biol 26:66–69

Préveral S, Gayet L, Moldes C, Hoffmann J, Mounicou S, Gruet A, … Forestier C (2009) A common highly conserved cadmium detoxification mechanism from bacteria to humans heavy metal tolerance conferred by the ATP-binding cassette (ABC) transporter SpHMT1 requires glutathione but not metal-chelating phytochelatin peptides. J Biol Chem 284(8):4936–4943

Rascio N, Navari-Izzo F (2011) Heavy metal hyperaccumulating plants: how and why do they do it? And what makes them so interesting? Plant Sci 180(2):169–181

Renault H, Bassard JE, Hamberger B, Werck-Reichhart D (2014) Cytochrome P450-mediated metabolic engineering: current progress and future challenges. Curr Opin Plant Biol 19:27–34

Rylott EL, Johnston EJ, Bruce NC (2015) Harnessing microbial gene pools to remediate persistent organic pollutants using genetically modified plants—a viable technology? J Exp Bot 66(21):6519–6533

Salvi S, Tuberosa R (2015) The crop QTLome comes of age. Cur Opin Biotech 32:179–185

Sarangi BK, Dash T, Pandey RA (2010) Engineering phytoremediation potentiality of plants through hyperaccumulation in plant biomass-with reference to arsenic and chromium. The Journal of Plant Science Research 26(2):113

Sayre R, Beeching JR, Cahoon EB (2011) The bio-cassava plus program: biofortification of cassava for Sub-Saharan Africa. Annu Rev Plant Physiol Plant Mol Biol 62:251–272

Seregin IV, Erlikh NT, Kozhevnikova AD (2014) Nickel and zinc accumulation capacities and tolerance to these metals in the excluder *Thlaspi arvense* and the hyperaccumulator *Noccaea caerulescens*. Rus J Plant Phys 61(2):204–214

Seth CS, Remans T, Keunen E, Jozefczak M, Gielen H, Opdenakker K, … Cuypers A (2012) Phytoextraction of toxic metals: a central role for glutathione. Plant Cell Environ 35(2):334–346

Singh P, Kumari S, Guldhe A, Misra R, Rawat I, Bux F (2016) Trends and novel strategies for enhancing lipid accumulation and quality in microalgae. Renew Sustain Energy Rev 55:1–16

Soric R, Leden can T, Zdunic Z et al (2012) Quantitative trait loci for metal accumulation in maize leaf. Maydica 56:323–327

Takahashi R, Ishimaru Y, Senoura T, Shimo H, Ishikawa S, Arao T, Nakanishi H, Nishizawa NK (2011) The OsNRAMP1 iron transporter is involved in Cd accumulation in rice. J Exp Bot 62:4843–4850

Takahashi R, Ishimaru Y, Shimo H, Ogo Y, Senoura T, Nishizawa NK, Nakanishi H (2012) The OsHMA2 transporter is involved in root-to-shoot translocation of Zn and Cd in rice. Plant Cell Environ 35(11):1948–1957

Tamayo E, Gomez-Gallego T, Azcon-Aguilar C, Ferrol N (2014) Genome-wide analysis of copper, iron and zinc transporters in the arbuscular mycorrhizal fungus Rhizophagus *irregularis*. Fron Plant Sci 5:547

Tan S, Han R, Li P, Yang G, Li S, Zhang P, Wang WB, Zhao WZ, Yin LP (2015) Over-expression of the Mx*IRT1* gene increases iron and zinc content in rice seeds. Trans Res 24:109–122

Tomar PR, Dixit AR, Jaiwal PK, Dhankher OP (2015) Engineered plants for heavy metals and metalloids tolerance. In: Genetic manipulation in plants for mitigation of climate change. Springer, New Delhi, pp 143–168

Tong YP, Kneer R, Zhu YG (2004) Vacuolar compartmentalization: a second-generation approach to engineering plants for phytoremediation. Trends Plant Sci 9(1):7–9

Uversky VN (2013) The alphabet of intrinsic disorder: II. various roles of glutamic acid in ordered and intrinsically disordered proteins. Intrin Dis Prot 1(1):e24684

Valls M, De Lorenzo V (2002) Exploiting the genetic and biochemical capacities of bacteria for the remediation of heavy metal pollution. FEMS Microbiol Rev 26(4):327–338

Vamerali T, Bandiera M, Mosca G (2010) Field crops for phytoremediation of metal contaminated land: a review. Environ Chemistry Lett 8:1–17

Wu Z, Banuelos GS, Lin ZQ, Liu Y, Yuan L, Yin X, Li M (2015) Biofortification and phytoremediation of selenium in China. Front Plant Sci 6:136

Xie P, Hao X, Herzberg M, Luo Y. Nies DH, Wei G (2015) Genomic analyses of metal resistance genes in three plant growth promoting bacteria of legume plants in Northwest mine tailings, China. J Environ Sci 27:179–187.

Xu N, Qian K, Dong Y, Chen Y, Yu Q, Zhang B, Li M (2014) Novel role of the *Candida albicans* ferric reductase gene CFL1 in iron acquisition, oxidative stress tolerance, morphogenesis and virulence. Res Microbio 165(3):252–261

Yadav SK (2010) Heavy metals toxicity in plants: an overview on the role of glutathione and phytochelatins in heavy metal stress tolerance of plants. South African Journal of Botany 76(2):167–179

Yadav R, Arora P, Kumar S, Chaudhury A (2010) Perspectives for genetic engineering of poplars for enhanced phytoremediation abilities. Ecotoxicology 19(8):1574–1588

Yan Z, Li X, Chen J, Tam NFY (2015) Combined toxicity of cadmium and copper in *Avicennia marina* seedlings and the regulation of exogenous jasmonic acid. Ecotoxic Environ Saf 113:124–132

Yang T, Chen ML, Wang JH (2015) Genetic and chemical modification of cells for selective separation and analysis of heavy metals of biological or environmental significance. Trend Anal Chem 66:90–102

Yu P, Yuan J, Zhang H, Deng X, Ma M, Zhang H (2016) Engineering metal-binding sites of bacterial CusF to enhance Zn/Cd accumulation and resistance by subcellular targeting. J Hazard Mater 302:275–285

Yuan L, Zhu Y, Lin ZQ, Banuelos G, Li W (2013) A novel selenocystine-accumulating plant in selenium-mine drainage area in Enshi, China. PLoS One 8:65615

Zhang Y et al (2013) Enhanced phytoremediation of mixed heavy metal (mercury)–organic pollutants (trichloroethylene) with transgenic alfalfa co-expressing glutathione S-transferase and human P450 2E1. J Hazard Mater 260:1100–1107

Zhao XQ, Wang RC, Lu XC, Lu JJ, Li J, Hu H (2012) Tolerance and biosorption of heavy metals by *Cupriavidus metallidurans* strain XXKD-1 isolated from a subsurface laneway in the Qixiashan Pb-Zn Sulfide Minery in Eastern China. Geomicro J 29(3):274–286

Zhao C, Xu J, Li Q, Li S, Wang P, Xiang F (2014) Cloning and characterization of a *Phragmites australis* phytochelatin synthase (PaPCS) and achieving Cd tolerance in tall fescue. PLoS One 9(8):103771

Zhou GK, Xu YF, Liu JY (2005) Characterization of a rice class II metallothionein gene: tissue expression patterns and induction in response to abiotic factors. J Plant Physiol 162(6):686–696

Zhou YH, Xue M, Yang ZY, Gong YL, Yuan JG, Zhou CY et al (2014) High cadmium pollution risk on vegetable amaranth and a selection for pollution-safe cultivars to lower the risk. Front Environ Sci Eng 7:219–230

Biodiversity Prospecting for Phytoremediation of Metals in the Environment

7

Contents

Abstract

Biodiversity prospecting would lead to the discovery of wild plants that could clean polluted environments of the world. This theme is at its infancy with a great anticipation for commercial. The flourishing monitoring methods for toxic metals in the environment are based on biosensors (microbe–metal interaction), i.e., gene- and protein-based biosensors. The fundamental aspects of microbe–plant stress responses to different doses of toxic metals together with breakthrough in biotechnology-based research innovations would successfully provide answers for application of biodiversity in advancement of phytoremediation technology. The appropriate plant selection vis-à-vis the phytoextraction is the key for bioprospecting.

Keywords

Biosensors • Ornamentals • Phytoremediation • Serpentinophytes

7.1 Introduction

Biodiversity prospecting leads to the discovery of wild plants for cleaning the environment possibly for the conservation of world's biodiversity by cleaning the contaminated environment. Though this concept is still at its infancy, the aspiration to

S. Jan, J.A. Parray, *Approaches to Heavy Metal Tolerance in Plants*,
DOI 10.1007/978-981-10-1693-6_7

capitalize on this new concept need to provide strong encouragement for conserving our biodiversity (De Groot et al. 2010). Toxic metals are increasing in all spheres of mother earth including atmosphere, biosphere, hydrosphere, and lithosphere on account of increasing anthropogenic pressure; likewise, the metal concentration in river waters and sediments increased many 1000-folds by effluents from industries (Ahmad and Goni 2010). The aquatic plants in all water systems are accumulators for most metals (Bo et al. 2015). In this direction, plant resources have been tested in field and laboratory for remediation that may be successful (Salomons and Forstner 2012).

The heavy metal monitoring in the environment is based on bacterial metal biosensors which are either gene-based or protein-based biosensors (Aksmann et al. 2014). Some lichens and erns are also capable of growing on metal-enriched substrates. These plants acquire anatomical and physiological characteristics enabling them to occupy specific ecological positions in natural metalliferous and man-made environment. For example, some taxonomic groups of bryophytes are seen on Cu-enriched substrates known as "copper mosses." And few are associated with Pb- and Zn-enriched substrates with little information is available about bryophytes growing on serpentine soils (Basile et al. 2008). However, pteridophytes (ferns) are reported to be associated with serpentine substrates, viz., brake fern, *Pteris vittata*, tolerates soil polluted with As of about 1500 ppm concentration, and its fronds concentrate the toxic metal to 22,630 ppm in 6 weeks (Ma et al. 2001). Around 400 metal hyperaccumulators from angiosperms have been notified that would serve as a reservoir for biotechnological purposes (Doronila 2012).

7.2 Metal Hyperaccumulators for Phytoremediation

In general, the mine reclamation and biogeochemical prospecting depends upon selection of specific plant species that is a reliable tool for achieving the goal of phytoremediation. It is reported that 163 plant taxa from 45 families are found to be metal tolerant and are capable of thriving on higher toxic metal concentrations. The right selection and utilizing the metal-tolerant species, indicator element, and accumulation power form the basis of biogeochemical prospecting (Ghosh and Singh 2005). It is mentioned that numerous Brassicaceae genera are well versed to accumulate heavy metals (Grispen et al. 2006). That is, about 11 genera and 87 species t are established for hyperaccumulation of toxic metals, and in particular Ni and Zn hyperaccumulation is reported in 7 genera and 72 species and 3 genera and 20 species, respectively (Grispen et al. 2006)., Zn.

Considerable progress had been achieved recently in unraveling the genetic secrets of metal-eating plants. Genes responsible for metal hyperaccumulation in plant tissues have been identified and cloned. These findings are expected to identify new nonconventional crops, *metallo-crops*, that can decontaminate metals in the environment (Jadia and Fulekar 2009). The basic aspects of microbes or plant stress responses to various doses of toxic metals together with insight through research novelty in biotechnology would endow us to utilize the plant genetic resources for advancing phytoremediation technology.

7.2.1 Ornamental Plants

Numerous studies report the potential role of numerous ornamental species for hyperaccumulation of toxic metals, viz., Nerium oleander leaves, in urban areas of Portugal accumulated about 78 mg/g of Pb dW and is appropriate for monitoring lead in air. Other plants like *Canna* x *generalis* are considered as potential Pb phytoextractors. In one study, *Pelargonium* sp. "Frensham," scented geranium under greenhouse conditions grown in artificial soil added with various metals, were capable of taking up good amounts of three major heavy metals, i.e., Pb, Ni, and Cd in a very short time (Fernandez and Rossini 2006). The plants were accomplished of extracting and accumulating in their root amounts of Pb, Cd, and Ni equivalent to 9 %, 2.7 %, and 1.9 % dw material, respectively, in 14 days. It was observed that within 10 years, the *Pelargonium* sp. could clean up heavily contaminated sites by toxic metals. The scientific studies indicate that growth and uptake and accumulation of metals in nutrient solution are extremely different to that in soil as hydroponic culture is not symptomatic of real-world circumstances, due to ion competition and root impedance, and may be because plants do not grow root hairs in solutions (Jadia and Fulekar 2009). For example, in phytoremediation, a Pb cleanup program that consists of 16 successive croppings of scented geranium plants at a density of 100 plants/sq meter could easily remove up to 72 g of Pbm^{-2} $years^{-1}$ (Mahdieh et al. 2013) Besides, the scented geranium also has the capability to endure on contaminated soils containing mixtures of toxic substances like heavy metals etc (Prasad 2003a).

7.2.2 Serpentinophytes and Metal Hyperaccumulation

Serpentinized rocks contain uneven distribution of some important elements, i.e., magnesium and calcium are distributed all over the world and harbors a distinct, often endemic, plant community (Harrison and Rajakaruna 2011). Serpentine soils are categorized by the presence of variable amounts of Mg in Ca and available Ni (Bani et al. 2009). The Ni toxicity in serpentine soils is due to high solubility in soil solution. Serpentine outcrops are mostly barrens because they are often characterized by scattered shrubby or stunted vegetation with small leathery leaves with extremely poor in essential nutrients. They are not of much agricultural value and can generally be distinguished by their gray-green or reddish-rocky soils. Serpentinophytes are regularly familiar with drought, nutrient stress, and excessive heavy metal exposure and high light intensity. The less availability of nutrients and water to plans in serpentine soils is due to the thin soil structure. However, deep serpentine soil particularly occurs only in valleys due to high leaching effects and enormous vegetation. In one study, the Northeast part of Portugal, the serpentinized area, accounts about 8000 ha with definite characteristic geology and vegetation. Besides serpentinophytes, some wild Brassicaceae are considered as best candidates for phytoremediation. Further, the serpentine habitats and species are threatened worldwide due to habitat loss, and many serpentine endemic species have become extinct or highly threatened. In this direction, our priority should be conservation of these regional hot spots of biodiversity (Marsili and Mariotti 2006).

7.2.3 Vegetables Crops

Various vegetable plants have been reported to have a property of metal accumulation. In one study, the dominant leaf vegetable producing plants, i.e., *Amaranthus spinosus*, *Alternanthera philoxeroides*, and *Alternanthera sessilis*, growing on the sewage sludge of Musi River were capable of bioaccumulation of some essential and nonessential elements like Cd, Zn, and Fe and whose concentrations were quite high in leaf tissues (Prasad and Freitas 2003), Pivetz (2001). *Alternanthera philoxeroides* was used for removal of Pb and Hg from contaminated water bodies. These edible plants would supplement the dietary requirement of human food with essential nutrients like Zn and Fe, but monitoring the metal transfer factor through food chain shall be taken care off (Gupta et al. 2013). So, the possibility of the use of these species to restore the biosolid and sewage sludge contaminated sites besides exercising caution on human consumption.

7.2.4 Plant Products as Biosorbents of Toxic Metals

A wide diversity of agricultural and forestry by-products are being used as biosorbents of heavy metals besides the normally used wild and cultivated plant species for the development of biofilters for precise applications. The number of plant by-products that are used to accumulate these toxic metals is depicted in Table 7.1.

Table 7.1 Biosorbents of toxic metals using plant by-products

Plant/agricultural products	Toxic heavy metals
Cotton, exhausted coffee grounds	Hg
Groundnut skin	Cu
Tree bark (*Pinus*, *Acacia*, etc.), freshwater green algae, apple waste, rice hulls, cellulose, and agrowaste	Cd, Cu, Pb, Ni, Cr, etc.
Waste tea leaves	Pb, Cd, and Zn
Pinus radiata	Uranium
Sawmill dust (wood waste)	Cr
Pinus pinaster bark	Zn, Cu, Pb
Sphagnum (moss peat), *Rhizopus arrhizus*	Cr(VI)
Aspergillus niger, *A. oryzae*	Cd, Cu, Pb, and Ni
Olive mill waste *Olea europaea*, marine algae	Cr, Pb, Cd, Zn, Cu, and Ni
Streptomyces rimosus, *Saccharomyces cerevisiae*	Cu and Ni
Aspergillus niger and *Rhizopus arrhizus*	Ni
Penicillium chrysogenum, *Fucus vesiculosus*, and *Ascophyllum nodosum* (marine algae)	Zn, Cu, and Ni
Phanerochaete chrysosporium, *P. versicolor*	Pb, Ni, Cr, Cd, Cu
Actionomycetes, *Aspergillus niger*, *A. oryzae*, *Rhizopus arrhizus*, *R. nigricans*	Cd
Immobilized *Pseudomonas putida* 5-X and *Aspergillus niger*, *Mucor rouxxi*	Cu
Rhizopus nigricans, *Phanerochaete chrysogenum*	Pb

Prasad and Freitas (2003), Dipu et al. (2011)

In one study, *Acacia nilotica* bark was reported to serve as an adsorbent of toxic metals. The bark (1 g) in 100 ml of aq. solution (10 mg ml^{-1} metal solution) exhibited different metal adsorption values of the order Cr>Ni>Cu>Cd>As>Pb. A similar trend of metal adsorption was observed for reused bark Cr>Ni>Cu>Cd>Pb—(Capuana 2011). Further, the results of toxicity bioassay confirmed the metal adsorption property of the bark powder as the leaf area; root length and number of new root primordia increased considerably in the presence of *A. nilotica* bark. It was concluded from the study that the metal budgets of the leaf and root confirmed the adsorbing property and, thus, lessen the toxicity imposed by the various tested toxic metals (Prasad and Freitas 2003). In another report, *Quercus ilex* L. phytomass from stem, leaf, and root acts as adsorbent of Cr, Ni, Cu, Cd, and Pb. The capacity of the metal uptake from roots was in the order of Ni>Cd>Pb>Cu>Cr, stem, Ni>Pb>Cu>Cd>Cr and leaf, and Ni>Cd>Cu>Pb>Cr. However, the highest amount adsorbed was Ni 7.30 nmol/g dw (root>leaf>stem) (Stoltz and Greger 2002). The results demonstrated that root biomass of the aforementioned plant has the ability of complexing Ni. Further, the results denote that the selected adsorbents are reusable (Prasad and Freitas 2003).

7.3 Elemental Allelopathy and Role of Hyperaccumulators and Serpentinophytes

Hyperaccumulators have the protective property against fungal and insect attacks, e.g., *Streptanthus polygaloides* and *Thlaspi montanum* are having Ni hyperaccumulation for the protection against fungal and bacterial pathogens (Dan et al. 2002; Mari et al. 2006). The hyperaccumulators and their associated properties on herbivory and influence of leachates on surrounding biota would add new height to the science of allelopathy. An antiherbivory effect of Zn has been found in the Zn hyperaccumulator *T. caerulescens* (Hu et al. 2009).

The metabolism of glutathione and organic acids plays a key role in metal tolerance in plants (Sun et al. 2005). Glutathione is a ubiquitous compound in organism cells, and its synthesis is mediated by the enzyme glutamylcysteine synthetase. However, low-molecular thiol peptides phytochelatins (PCs) often called class III metallothioneins are synthetized in plants from glutathione induced by heavy metals (Kramer 2010). The main precursor of these peptides is called phytochelatin synthase (PCS). The PCS is a cytosolic, constitutive enzyme which is activated by various metal ions, i.e., Hg^{2+}, Zn^{2+}, Cu^{2+}, Cd^{2+}, Pb^{2+}, Ag^{1+}, Bi^{3+}, and Au^{2+}. Thus, these metal complexes are transported through cytosol in an ATP-dependent manner through tonoplast into vacuole (Huang et al. 2010). In this regard, toxic metals are removed from cytosol (Clemens and Persoh 2009). Under heavy metal stress, plants produce free radicals and ROS for withstanding oxidative stress besides tolerance to toxic elements, and in plants, the ascorbate-glutathione pathway is involved for plant defenses. Meanwhile, the transgenic plants with modified genes of PCS and glutathione synthesis enzymes should be the primary concern for achieving the success in the field of phytoremediation (Liang et al. 1999; Pilon-Smits et al. 1999). Organic acids also play a major role in metal tolerance in plants (Ma et al. 2001) by forming complexes with toxic metals, by a process known as metal detoxification. Metal chelation with organic acids in the rhizosphere and its various processes are indeed an important aspect of phytoremediational study. All these metabolic pathways highlight the physiological, biochemical, and molecular basis for toxic metal tolerance in plants (Prasad 2003a, b).

7.4 Molecular and Transgenic Approaches for Phytoremediation

Transgenics and other genetic approaches in plants and microbial production vis-à-vis the field trials are an important attributes for phytoremediation (Barabasz et al. 2012). Hg is a global problem as an outcome of its different uses in bleaching operations (chlorine production, textiles, etc.) and besides used as a catalyst, pigment for paints, gold mining, and fungicides. Elemental mercury, Hg (0), and its other forms, methylmercury (CH_3-Hg^+), dimethylmercury $(CH_3)_2$-Hg), or other organomercury salts, can accumulate in animals or higher organisms known as biomagnifications responsible for severe neurological degeneration in birds, cats, and humans (Meagher et al. 2000). However, some microbes have an ability of pumping toxic metals out of their cell or by converting them into less toxic metals. For example, mer operon contains different genes that sense mercury (MerB), transport mercury (merT), sequester mercury to the periplasmic space (merP), and reduce mercury (merA). The MerB is a subset of the mer operon which encodes an enzyme, organomercurial lyase, and catalyzes the breakdown of various forms of organic mercury to Hg^{2+}(Pilon-Smits and Pilon 2000).

7.5 Phytoremediation Technology for Enhancing Chelation

Soil amendments like synthetics (ammonium thiocyanate) and natural zeolites have yielded overwhelming results so far. Numerous substances like EDTA, citrate, oxalate, malate, succinate, tartrate, salicylate, acetate, etc. have been used as chelators for speedy mobility and metal uptake from contaminated soils by different plant species (Shahid et al. 2014). The use of synthetic chelators significantly facilitates the phytoextraction of some important toxic metals like Pb and Cd from contaminated sites via uptake and translocation from roots to other plant parts. Some synthetic polyacrylates or hydrogels protect plant roots from metal toxicity, and entry of metals is also restricted into roots. The low-cost synthetic compounds and natural zeolites are applied on large scale to the soil via irrigation at definite stages of plant growth that is productive to accelerate heavy metal accumulation (Pociecha and Lestan 2010).

The efficiency of phytoextraction by plants depends upon the time period in which sufficient metals are absorbed in very shorter period of time. Hyperaccumulators accumulate considerable quantities of heavy metals in their tissue despite the concentration of metals in the soil, contrasting moderate accumulators where the quantity of absorbed metals is an indication of their concentrations in the soil and besides also depends upon the availability and the efficiency of metal absorption by plant roots. To speed up the efficiency for uptake of metals, the use of various chemicals substances have been advocated which includes various acidifying agents (Chen et al. 2012), fertilizer salts, and other chelating materials (Cui et al. 2007). They either help in liberating or displacing toxic metals from soil or by leach out metals from complexes. Among the physical factors, pH of soil is an important regulating factor that influences uptake of metals by plants. The H^+ions dislocate cations from the cation exchange complex and cause metals to be released from sesquioxides and variable-charged clays to which they have been chemisorbed (i.e., specific adsorption) in soil under acidic conditions. The phytoextraction process is elevated by plant roots via addition of acidifying agents like NH_4-containing fertilizers, organic and inorganic acids, and sulfur to the soil. Various metal cations like Cd, Cu, Hg, Ni, Pb, and Zn have been reported to be more available in the soil solution at pH below 5.5 (Peng et al. 2009) (Blaylock and Huang 2000). Many reports (Shahid et al. 2014) suggest the use of elemental S for acidification of soil and are common agronomic practice, for mobilization of metal cations in soils. For example, CZn- and Cd-contaminated soils were acidified with elemental S, and accumulation of metals by plants was found higher. Significantly, these acidifying agents increase the solubility vis-à-vis the availability of radioactive elements like Jean et al. (2012) who noted that by addition of citric acid, uranium accumulation in *Brassica juncea* increases than other agents used. This might be due to the chelating property of citric acid, which causes 1000-fold increase of U uptake by *B. juncea* roots. In this context, more studies need to be taken for the phytoextraction of heavy or toxic metals from contaminated soil.

The chelating substances such as EDTA, HEDTA, and EDDHA are most effective for liberating metal contaminants into soil solution. The property of chelates to form complexes allows dissolution of the sorbed or precipitated compounds for easy accumulation of metal ions (Neugschwandtner et al. 2008). They are used for enhancing the phytoextraction of various toxic metals like Cd, Cu, Ni, Pb, and Zn (Grcman et al. 2001). The chelates increase Pb accumulation of 1000-fold increase in some agronomic crops, i.e., *Zea mays* L. and *Pisum sativum* L. after HEDTA application (Murakami et al. 2008). The accumulation of toxic metals in nonaccumulating plant species is known as chelate-induced hyperaccumulation (Evangelou et al. 2007). Chelate-induced hyperaccumulation allows more metal accumulation over time as the plants grow (Hsiao et al. 2007). More data is not available pertaining the suitable amount of chelates for applying under different contamination levels and also for variable plant species. Further, more research work needs to be carried out to determine the fate of these chelate–metal complexes in soil before they are widely used for phytoextraction (Neugschwandtner et al. 2008).

The ultimate success of phytoextraction is dependent upon proper plant selection with fast-growing and higher accumulation ability (Li et al. 2009). Around 400 known hyperaccumulators of plant species have been reported so far (Sun et al. 2009), but still most are not suitable for phytoextraction because of their slow growth (Odjegba and Fasidi 2004; Yanai et al. 2006; Wu et al. 2009; Yuan et al. 2015). Some metal-tolerant plants like grasses and flee toxicity through an exclusion mechanism are therefore suitable for phytostabilization than phytoextraction. Herbaceous species, including the members of family Brassicaceae, viz., Indian mustard (*B. juncea*) also accumulate reasonable amounts of a range of toxic metals in their shoots (Bernard et al. 2015). In one report, Das et al. (2011) studied different fast-growing Brassicas for their ability to tolerate and accumulate toxic metals, like Indian mustard (*B. juncea*), black mustard (*Brassica nigra* Koch), turnip (*Brassica campestris* L.), rape (*Brassica napus* L.), and kale (*Brassica oleracea* L). It was observed that *B. juncea* showed a higher ability for accumulation and translocation of metals like Cu, Cr (VI), Cd, Ni, Pb, and Zn than other Brassicas. The biomass accumulation *of B. juncea* was greatest in soil supplied with 200:100:66 of N, $P\ _2O_5$, and K_2O kg/hac. The Cd concentration in the tissues was higher under no N supplement (Armas et al. 2015). Further, the USDA-ARS Plant Introduction Station maintains a worldwide collection of *B. juncea* accessions that are known metal accumulators, and the seeds are distributed to public and private research institutions at no cost (Sikka and Nayyar 2012).

References

Ahmad JU, Goni MA (2010) Heavy metal contamination in water, soil, and vegetables of the industrial areas in Dhaka, Bangladesh. Environ Monit Assess 166:347–357

Aksmann A, Pokora W, Bascik-Remisiewicz A, Dettlaff-Pokora A, Wielgomas B, Dziadziusko M (2014) Time-dependent changes in antioxidative enzyme expression and photosynthetic activity of *Chlamydomonas reinhardtii* cells under acute exposure to cadmium and anthracene. Ecotoxicol Environ Saf 110:31–40

Armas T, Pinto AP, de Varennes A, Mourato MP, Martins LL, Gonçalves MLS, Mota AM (2015) Comparison of cadmium-induced oxidative stress in *Brassica juncea* in soil and hydroponic cultures. Plant Soil 388:294–305

Bani A, Echevarria G, Mullaj A, Reeves RD, Morel JL, Sulçe S (2009) Ni hyperaccumulation by *Brassicaceae* in serpentine soils of Albania and NW Greece. Northeast Nat 16(5):385–404

Barabasz A, Wilkowska A, Ruszczy A, Ska A, Bulska E, Hanikenne M (2012) Metal response of transgenic tomato plants expressing P(1B) -ATPase. Physiol Plant 145:315–331

Basile A, Sorbo S, Aprile G, Conte B, Castaldo Cobianchi R (2008) Comparison of the heavy metal bioaccumulation capacity of an epiphytic moss and an epiphytic lichen. Environ Pollut 151(2):401–407

Bernard F, Brulle F, Dumez S, Lemiere S, Platel A, Nesslany F, Cuny D, Deram A, Vandenbulcke F (2015) Antioxidant responses of annelids, Brassicaceae and fabaceae to pollutants: a review. Ecotoxicol Environ Saf 114:273–303

Blaylock MJ, Huang JW (2000) Phytoextraction of metals. In: Raskin I, Ensley BD (eds) Phytoremediation of toxic metals: using plants to clean up the environment. John Wiley and Sons, Inc, Toronto, p 303

Bo L, Wang D, Li T, Li Y, Zhang G, Wang C (2015) Accumulation and risk assessment of heavy metals in water, sediments, and aquatic organisms in rural rivers in the Taihu Lake region. China Environ Sci Pollut Res 22:6721–6731

Capuana M (2011) Heavy metals and woody plants—biotechnologies for phytoremediation. Biogeosci Forestry 4:7–15

Chen BC, Lai HY, Juang KW (2012) Model evaluation of plant metal content and biomass yield for the phytoextraction of heavy metals by switchgrass. Ecotoxicol Environ Saf 80:393–400

Clemens S, Persoh D (2009) Multi-tasking phytochelatin synthases. Plant Sci 177:266–271

Cui S, Zhou Q, Wei S, Zhang W, Cao L, Ren L (2007) Effects of exogenous chelators on phytoavailability and toxicity of Pb in *Zinnia elegans* Jacq. J Hazard Mater 146:341–346

Dan TV, Krishnaraj S, Saxena PK (2002) Cadmiumand nickel uptake and accumulation in scented geranium (*Pelargonium* sp. frensham). Water Air Soil Pollut 137:355–364

Das S, Sen M, Saha C, Chakraborty D, Das A, Banerjee M, Seal A (2011) Isolation and expression analysis of partial sequences of heavy metal transporters from *Brassica juncea* by coupling high throughput cloning with a molecular fingerprinting technique. Planta 23:139–156

De Groot RS, Alkemade R, Braat LC, Hein L, Willemen L (2010) Challenges in integrating the concept of ecosystem services and values in landscape planning, management and decision making. J Ecol Complex 7(3):260–27

Dipu S, Anju AK, Salom Gnana Thanga V (2011) Phytoremediation of dairy effluent by constructed wetland technology. *Environmentalist*, 31(3):263–268

Doronila AI (2012) Clean and green: plants as metal prospectors in the Philippines. Chem Aust 3:26–28

Evangelou MWH, Ebel M, Schaeffer A (2007) Chelate assisted phytoextraction of heavy metals from soil. Effect, mechanism, toxicity, and fate of chelating agents. Chemosphere 68:989–1003

Fernandez E, Rossini AJ (2006) The composition and relation-ships between trace element levels in inhalable atmospheric particles (PM10) and in leaves of Nerium oleander L. and Lantana camara L. Chemosphere 62:1665–1672

Ghosh M, Singh SP (2005) A review on phytoremediation of heavy metals and utilization of it's by products. Appl Ecol Environ Res 3(1):1–18

Grcman H, Velikonja-Bolta S, Vodnik D, Kos B, Lestan D (2001) EDTA enhanced heavy metal phytoextraction: metal accumulation, leaching and toxicity. Plant Soil 235:105–114

Grispen VMJ, Nelissen HJM, Verkleij JAC (2006) Phytoextraction with *Brassica napus* L.: a tool for sustainable management of heavy metal contaminated soils. Environ Pollut 144(1):77–83

Gupta S, Jena VS, Jena N, Davic N, Matic D, Radojevic SJS (2013) Assessment of heavy metal contents of green leafy vegetables. Croat J Food Sci Technol 5:53–60

Harrison S, Rajakaruna N (2011) Serpentine: evolution and ecology in a model system. University of California Press, Berkely

Hsiao KH, Kao PH, Hseu ZY (2007) Effects of chelators on chromium and nickel uptake by *Brassica juncea* on serpentine mine-tailings for phytoextraction. J Hazard Mat 148:366–376

Hu PJ, Qiu RL, Senthilkumar P, Jiang D, Chen ZW, Tang YT, Liu FJ (2009) Tolerance, accumulation and distribution of zinc and cadmium in hyperaccumulator *Potentilla griffithii*. Environ Exp Bot 66:317–325

Huang GY, Wang YS, Sun CC, Dong JD, Sun ZX (2010) The effect of multiple heavy metals on ascorbate, glutathione and related enzymes in two mangrove plant seedlings (*Kandelia candel* and *Bruguiera gymnorrhiza*). Oceanol Hydrobiol Stud 39(1):11–25

Jadia CD, Fulekar MH (2009) Phytoremediation of heavy metals: recent techniques. Afr J Biotechnol 8(6):921–928

Jean SL, Bordas F, Bollinger JC (2012) Column leaching of chromium and nickel from a contaminated soil using EDTA and citric acid. Environ Pollut 164:175–181

Kramer U (2010) Metal hyperaccumulation in plants. Annu Rev Plant Biol 61:517–534

Li JT, Liao B, Dai Z, Zhu R, Shu WS (2009) Phytoextraction of Cd-contaminated soil by carambola (*Averrhoa carambola*) in field trials. Chemosphere 76(9):1233–1239

Liang YC (1999) Effects of silicon on enzyme activity, and sodium, potassium and calcium concentration in barley under salt stress. Plant Soil 209:217–224

Ma LQ, Komar KM, Tu C, Zhang W, Cai Y, Kennelley ED (2001) A fern that hyperaccumulates arsenic. Nature 409:579–579

Mahdieh M, Yazdani M, Mahdieh S (2013) The high potential of Pelargonium roseum plant for phytoremediation of heavy metals. Environ Monit Assess 185(9):7877–7881

Mari SD, Gendre K, Pianelli L, Ouerdane LR (2006) Root-to-shoot long-distance circulation of nicotianamine and nicotianamine-nickel chelates in the metal hyperaccumulator *Thlaspi caerulescens*. J Exp Bot 57:4111–4122

Marsili S, Mariotti M (2006) The European Habitat Directive and the conservation of the nature in the ophiolitic Areas. In: Fifth international conference on serpentine ecology, Abstract Book. University of Siena, Italy

Meagher RB, Rugh CL, Kandasamy MK, Gragson G, Wang NJ (2000) Engineered phytoremediation of mercury pollution in soil and water using bacterial genes. In: Terry N, Banuelos G (eds) Phytoremediation of contaminated soil and water. Lewis, Boca Raton, pp 201–221

Murakami M, Nakajima F, Furumai H (2008) The sorption of heavy metal species by sediments in soakaways receiving urban road runoff. Chemosphere 70(11):2099–2109

Neugschwandtner RW, Tlustos P, Komárek M, Száková J (2008) Phytoextraction of Pb and Cd from a contaminated agricultural soil using different EDTA application regimes: laboratory versus field scale measures of efficiency. Geoderma 144(3–4):446–454

Odjegba VJ, Fasidi IO (2004) Accumulation of trace elements by *Pistia stratiotes*: implications for phytoremediation. Ecotoxicology 13(7):637–646

Peng JF, Song YH, Yuan P, Cui XY, Qiu GL (2009) The remediation of heavy metals contaminated sediment. J Hazard Mater 161:633–640

Pilon-Smits EAH, Pilon M (2000) Breeding mercury-breathing plants for environmental cleanup. Trends Plant Sci 5:235–236

Pilon-smits EAH, Hwang S, Lytle CM, Zhu YL et al (1999) Overexpression of ATP sulfurylase in Indian mustard leads to increased selenite uptake, reduction, and tolerance. Plant Physiol 119(1):123–132

Pivetz BE (2001) Ground water issue: phytoremediation of contaminated soil and ground water at hazardous waste sites. ManTech Environmental Research Services Corporation, Ada, pp 1–36

Pociecha M, Lestan D (2010) Electrochemical EDTA recycling with sacrificial Al anode for remediation of Pb contaminated soil. Environ Pollut 158:2710–2715

Prasad MNV (2003a) Phytoremediation of metal-polluted ecosystems: hope for commercialization. Russ J Plant Physiol 50:764–780

Prasad MNV (2003b) Metal hyperaccumulation in plants—biodiversity prospecting for phytoremediation technology. Electron J Biotechnol 6(3):0717–3458

Prasad MNV, Freitas HMO (2003) Metal hyperaccumulation in plants—biodiversity prospecting for phytoremediation technology. Electron J Biotechnol 6(3):285–321

Salomons W, Förstner U (2012) Metals in the hydrocycle. Springer Science & Business Media, Berlin/Heidelberg

Shahid M, Austruy A, Echevarria G, Arshad M, Sanaullah M, Aslam M, Nadeem M, Nasim W, Dumat C (2014) EDTA-enhanced phytoremediation of heavy metals: a review. Soil Sediment Contam 23:389–416

Sikka R, Nayyar V (2012) Cadmium accumulation and its effects on uptake of micronutrients in Indian mustard [*Brassica juncea* (L.) czern.] grown in a loamy sand soil artificially contaminated with cadmium. Commun Soil Sci Plant Anal 43:672–688

Stoltz E, Greger M (2002) Accumulation properties of As, Cd, Cu, Pb and Zn by four wetland plant species growing on submerged mine tailings. Environ Exptal Botany 47:271–280

Sun Q, Ye ZH, Wang XR, Wong MH (2005) Increase of glutathione in mine population of sedum alfredii: a Zn hyperaccumulator and Pb accumulator. Phytochemistry 66:2549–2556

Sun Y, Zhou Q, Wang L, Liu W (2009) Cadmium tolerance and accumulation characteristics of *Bidens pilosa* L. as a potential Cd-hyperaccumulator. J Hazard Mater 161(2–3):808–814

Wu F, Yang W, Zhang J, Zhou L (2009) Cadmium accumulation and growth responses of a poplar (*Populus deltoids* × *Populus nigra*) in cadmium contaminated purple soil and alluvial soil. J Hazard Mater 177:268–273

Yanai J, Zhao FJ, McGrath SP, Kosaki T (2006) Effect of soil characteristics on Cd uptake by the hyperaccumulator *Thlaspi caerulescens*. Environ Pollut 139(1):167–175

Yuan H, Liu E, Shen J (2015) The accumulation and potential ecological risk of heavy metals in microalgae from a eutrophic lake (Taihu Lake, China). Environ Sci Pollut Res 22:17123–17134

MIX
Papier aus verantwortungsvollen Quellen
Paper from responsible sources
FSC® C105338

If you have any concerns about our products,
you can contact us on
ProductSafety@springernature.com

In case Publisher is established outside the EU,
the EU authorized representative is:
Springer Nature Customer Service Center GmbH
Europaplatz 3, 69115 Heidelberg, Germany

Printed by Libri Plureos GmbH
in Hamburg, Germany